2nd International Symposium on Hydrogen Energy and Energy Technologies (HEET 2019)

E3S Web of Conferences Volume 155 (2020)

Osaka, Japan
12 - 13 December 2019

Editor:

Yun Guo

ISBN: 978-1-7138-0823-7

Printed from e-media with permission by:

Curran Associates, Inc.
57 Morehouse Lane
Red Hook, NY 12571

Some format issues inherent in the e-media version may also appear in this print version.

For additional information, please contact EDP Sciences – Web of Conferences at the address below.

EDP Sciences – Web of Conferences
17, Avenue du Hoggar
Parc d'Activité de Courtabœuf
BP 112
F-91944 Les Ulis Cedex A
France

Phone: +33 (0) 1 69 18 75 75
Fax: +33 (0) 1 69 28 84 91

contact-edps@webofconferences.org

Additional copies of this publication are available from:

Curran Associates, Inc.
57 Morehouse Lane
Red Hook, NY 12571 USA
Phone: 845-758-0400
Fax: 845-758-2633
Email: curran@proceedings.com
Web: www.proceedings.com

TABLE OF CONTENTS

Preface

This is the proceeding of the second International Symposium on on Hydrogen Energy and Energy Technologies (HEET 2019) held at ANA CROWNE PLAZA OSAKA, Japan during December12-13, 2019.

The vision of the HEET 2019 is a premier interdisciplinary platform for researchers, practitioners and educators to present and discuss the most recent innovations, trends, and concerns as well as practical challenges encountered and solutions adopted in the fields of Hydrogen Energy & Energy Technologies. The conference has successfully performing forum to transferring and discussing research result among the researcher, students, government, private sector or industries.

We are thankful to all the authors who submitted papers and the delegates for their participation and their interest in HEET as a platform to share their ideas and innovation. Also, we are also thankful to all the committee members for providing continuous guidance and support and Efforts taken by peer reviewers contributed to improve the quality of papers provided constructive critical comments, improvements and corrections to the authors are gratefully appreciated for their contribution to the success of the conference.

edp sciences

Web of Conferences

Statement of Peer review

In submitting conference proceedings to *Web of Conferences*, the editors of the proceedings certify to the Publisher that

1. They adhere to its **Policy on Publishing Integrity** in order to safeguard good scientific practice in publishing.
2. All articles have been subjected to peer review administered by the proceedings editors.
3. Reviews have been conducted by expert referees, who have been requested to provide unbiased and constructive comments aimed, whenever possible, at improving the work.
4. Proceedings editors have taken all reasonable steps to ensure the quality of the materials they publish and their decision to accept or reject a paper for publication has been based only on the merits of the work and the relevance to the journal.

Title, date and place of the conference

The 2nd International Symposium on Hydrogen Energy and Energy Technologies (HEET 2019)

December 12-13, 2019 / Osaka, Japan

Proceedings editor(s): Yun Guo

Date and editor's signature

2020/2/28

Hydrogen pump for hydrogen recirculation in fuel cell vehicles

Wilhelm Wiebe[1]*, Thomas v. Unwerth[2] and Sven Schmitz[3]

[1] M.Eng. Wilhelm Wiebe, DHBW Mannheim
[2] Prof. Dr.-Ing. Thomas von Unwerth, Technische Universität Chemnitz
[3] Prof. Dr.-Ing. Sven Schmitz, DHBW Mannheim

Abstract. A promising alternative to fossil-fuelled vehicles are battery-powered vehicles and fuel cell (FC) vehicles. The major differences between fuel cell and battery-powered vehicles are the range and refuelling times of each vehicle type. With a hydrogen (H_2) fuelling time of approx. 5 minutes it is possible to cover a distance of up to 800 km with a fuel cell vehicle. These properties make a fuel cell vehicle comparable to a fossil fuel powered vehicle. Furthermore, due to short fuelling times and long range capabilities, fuel cell vehicles are more suitable for long-distance, trucking and agriculture than battery-powered vehicles. The aim of current research is to increase the profitability of fuel cells by reducing costs and improving performance. To ensure a high performance of the fuel cell stack, more hydrogen is supplied to the stack than is needed for the reaction. Therefore, unused hydrogen is pumped back to the anode inlet of the FC-stack using a jet pump or a recirculation blower. In this study, the application of an electrochemical compressor or hydrogen pump (HP) for hydrogen recirculation is suggested. The hydrogen pump is an innovative H_2 transport technology with the additional functions of compression and purification in the recirculation system. Hydrogen pumps are very efficient compared to mechanical compressors due to the almost isothermal conditions they operate under. Furthermore, due to the modular design, hydrogen compressors can utilize a minimal amount of space in vehicles.

Keywords: Hydrogen pump, electrochemical compressor, fuel cell, hydrogen recirculation, hydrogen purification, hydrogen efficiency, electric power consumption, efficiency of electric drives.

1. Introduction

1.1 Short Description of the Necessity for Recirculation

In the operation of a fuel cell hydrogen is typically supplied in excess to prevent local hydrogen starvation on the membrane electrode assembly (MEA) and to increase the life of the fuel cell. The hydrogen ratio $\boldsymbol{\lambda}_{H_2}$ can be calculated using Eq. (1),

$$\lambda_{H_2} = \dot{n}_{H_2,FC,in} / \dot{n}_{H_2,FC,use} \,. \tag{1}$$

With the use of a passive jet pump, an active recirculation blower, or a combination of both, the unused hydrogen is pumped back to the anode inlet and combined with fresh hydrogen from the tank (Fig.1).

* Corresponding author: wwiebe@dhbw-mannheim.de

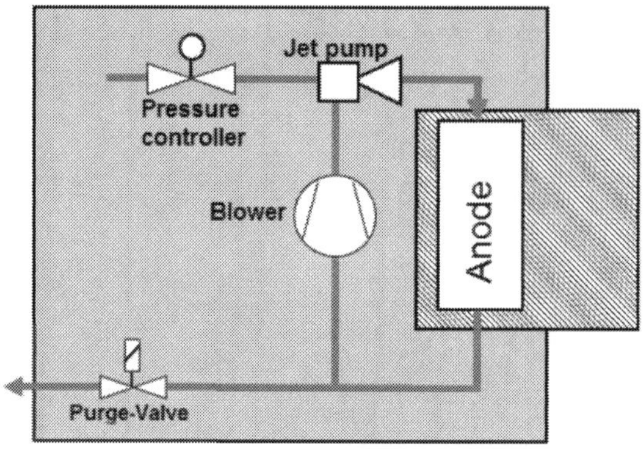

Fig.1. Hydrogen recirculation with jet pump and blower

Due to the diffusion and permeation through the membrane, a parasitic exchange of gases takes place (Fig.2). Hydrogen and oxygen oxidize to water vapour, causing additional humidification of the MEA.

Nitrogen permeation through the membrane is responsible for a concentration of inert gases on the anode side and prevents a uniform reaction from occurring. A less uniform reaction due to the presence of nitrogen inherently causes a decrease in FC performance. For this reason, the anode gas is released at regular intervals with a purge-valve to remove the nitrogen from the system (Fig.1). However, purging also causes the loss of hydrogen, lowering the efficiency of the system.

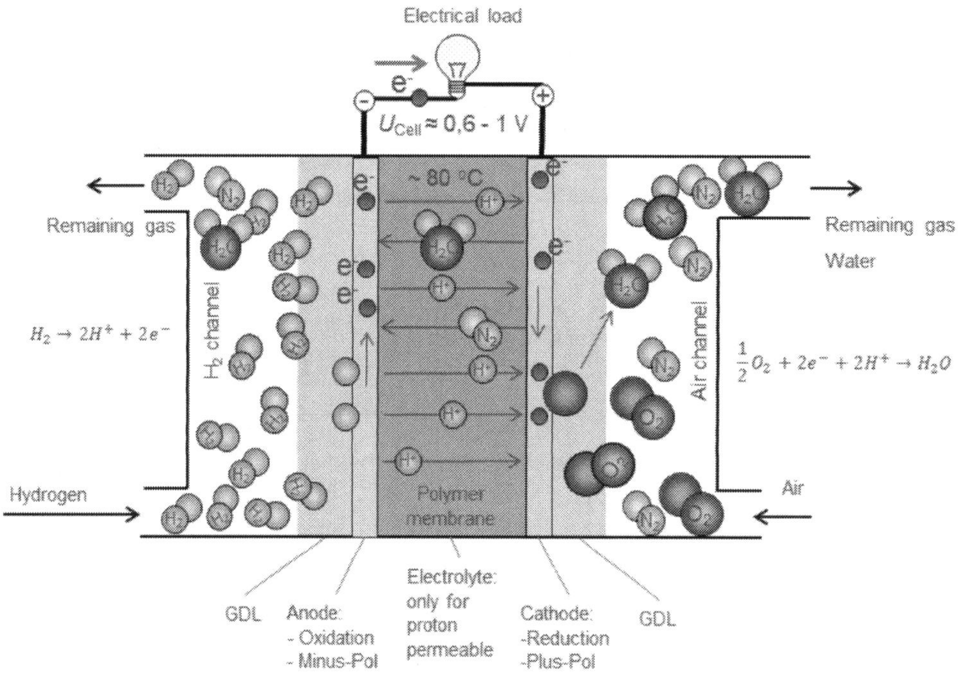

Fig.2. Gas diffusion through the FC-membrane

1.2 Replacement of the Blower by a Hydrogen Pump

This study proposes a **new option** for recirculating hydrogen in a fuel cell system in the form of an electrochemical compressor. The electrochemical compressor, also called hydrogen pump (HP) may be a modified polymer electrolyte membrane fuel cell which is connected to a constant current electrical current source for pumping hydrogen. Thus, there is no electrochemical energy conversion as in a fuel cell, but a hydrogen transport and an increase in pressure without mechanically moving components.

1.3 Function of a Hydrogen Pump

The structure of the electrochemical compressor is very similar to a FC. Like a fuel cell, the HP consists of an anode, a cathode and an electrolyte. The electrolyte of the HP is a polymer electrolyte membrane (PEM). By applying an electrical potential from an external constant current source, the hydrogen is forced to split up into protons and electrons. The difference in potential leads to a transport of the protons to the cathode side through the membrane. Contrary to the fuel cell, however, no gas is supplied to the electrochemical compressor on the cathode side. Thus, no reaction between hydrogen and oxygen is possible. At the cathode side, the electrons and protons recombine again to hydrogen and continuous pumping increases the pressure on the cathode side in a closed system.
Fig.3 shows how the hydrogen pumping action separates gases as only hydrogen protons are pumped through the membrane. Other gases such as nitrogen are unaffected by the pumping action, and remain on the anode side where they can be purged. Other gases can only transport from the anode to the cathode side of the HP by diffusion processes.
When integrating a hydrogen pump with polymer membrane (PEM-HP) into a fuel cell system, the electrochemical compression has the following advantages compared to the mechanical compression:

- The recirculated hydrogen is purified from a hydrogen-containing gas mixture and passed to the FC, thereby increasing the efficiency of the FC.
- All contamination in the anode gas is purged and the anode purge flow of the HP contains little or no amounts of hydrogen, minimizing the loss of the hydrogen from the tank.
- In nearly isothermal conditions, the HP has a higher efficiency compared to mechanical compressors.
- HPs are sturdy and have low maintenance costs because they contain no moving parts. Compared to noisy mechanical compressors, there is no risk of contamination of the hydrogen from oil particles and abrasion.
- HPs require the same components of fuel cells, making them modular and simple to integrate into a fuel cell stack (anode, cathode, and electrolyte). Therefore, and HP and fuel cell system requires significantly less space than mechanical alternatives to HPs.

The main functions of the HP in the proposed application are the separation of hydrogen from the fuel cell exhaust mixture and the pumping of the extracted hydrogen [1]. In the literature, other possible applications for the hydrogen pump are proposed such as the

- hydrogen compression for storage [2],
- hydrogen purification [3, 4].

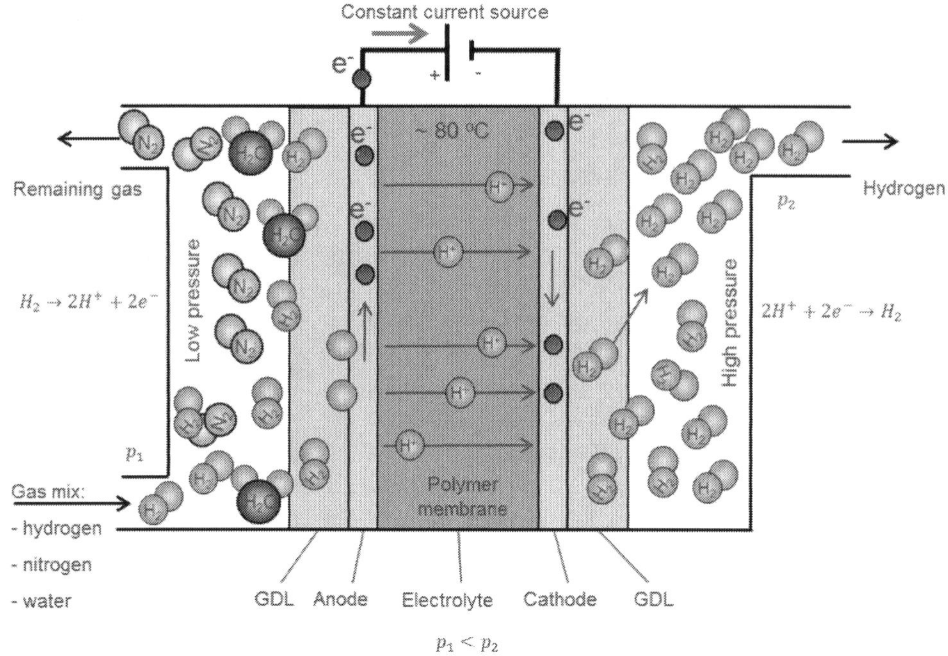

Fig.3. Schematic overview of a hydrogen pump

2 Experimental and Numerical Methods

2.1 Simulation

Within the scope of this work simulations were performed using Comsol Multiphysics. The aim of this work is the theoretical design and optimization of a hydrogen pump for hydrogen recirculation in an automotive fuel cell system.
With the findings from the simulation, the potential of the HP should be more accurately estimated and can be validated through experimentation.
The simulation includes a complete fuel cell system with a HP with an active area of 20 cm² and considers all electrochemical processes. The focus of the simulations is the mass transport through the membrane of the HP. The gases involved and their transport processes are examined closely and a prediction for the scaling of the PEM-HP to typical effective areas in fuel cell vehicles of 320 to 400 cm² is made.

2.2 Experiment

The theoretical relationship between the current density and the resulting volume was experimentally tested to determine if the relationship follows the Faraday law without any restrictions.
For the experiments, Umicore pMembrain H300 MEA's with an active membrane area of 100 cm² and a membrane thickness of 25 µm were used. Carbon electrodes with a platinum loading of 0.4 mg/cm² were applied directly on each side of the membrane as the anode and cathode. A Sigracet GDL 10 BB gas diffusion layer (GDL) with a slightly larger dimension of 103 cm² was also applied directly to the electrodes. The flow distributor for the hydrogen and oxygen side consists of two laser-welded stainless steel sheets, each embossed with an 8-fold meander. The interspace of the welded sheets was used for the transport of demineralized water for the cooling of the cell.

Fig.4 shows the schematic structure of the HP in the test bench. A gas mixture of 60 % H_2 and 40 % N_2 was supplied from the gas tank. After setting the flow rate by a mass flow controller, humidified hydrogen was fed to the anode inlet. The pressures and temperatures as well as the H_2 mole fraction were also measured.
Several experiments were carried out with a gas mixture of 60 % H_2 and 40 % N_2 at different temperatures (50-90 °C) and relative humidity (60-90 % r.H.).
The cell temperature was initially set to 60 °C and relative humidity was set to 60 %. The operating pressure was 1.5 bar at the HP anode side and 1.6 bar at the cathode side.
The hydrogen pump was operated at a constant flow rate and the voltage-current characteristics were recorded.
On the cathode side, the relative humidity transferred by electroosmotic drag was first measured. After the separator, the flow rate, temperature, and H_2-mole fraction of dry hydrogen were determined. Subsequently, the flow rates from the experiment were compared with the theoretical flow rates according to Faraday's law with direct dependence on the current. In the next step, the series of measurements with higher temperatures and/or humidities were repeated. In parallel, reference cells have been taken from the literature and have been compared with the experimental results.

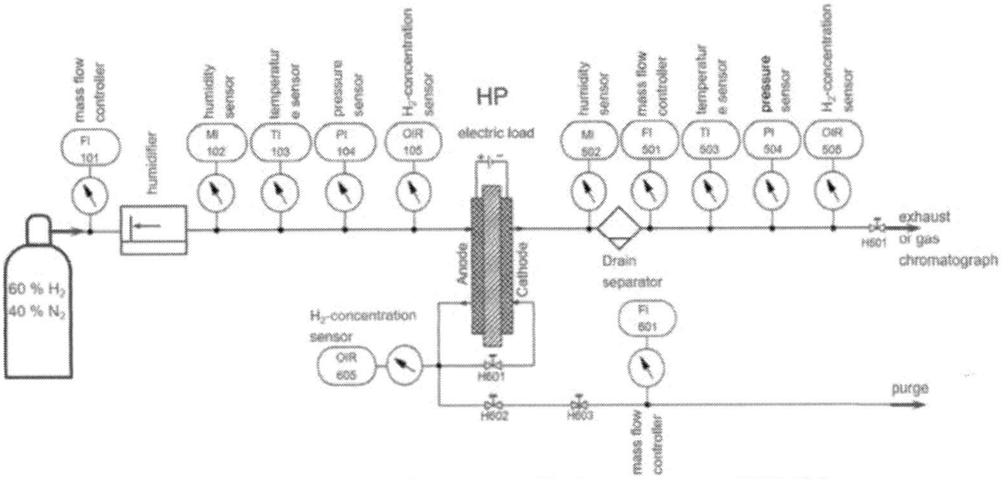

Fig.4. Schematic diagram of experimental apparatus of hydrogen pump and PEMFC.

3 Results and Discussion

3.1 Inert Gas Discharge

The aim of this study was to separate the hydrogen from the hydrogen mixture with a gas composition of 60 % H_2 and 40 % N_2, to examine the hydrogen purity at the HP cathode output. The cleaning effect by the HP was confirmed in the experiments. After operation of the HP, the concentration of the gas mixture on the cathode side was ~99.5 % H_2 and only ~ 0.5 % of nitrogen (Fig.5).

In the experimental studies with a gas composition between 2 and 5 % N_2 at the anode inlet, which are similar to the realistic nitrogen content in the vehicles, the nitrogen content at the cathode was so small that the sensors were not able to detect it. Further examination of the cathode gas content with a gas chromatograph indicated a nitrogen content of ~ 20-40 ppm.

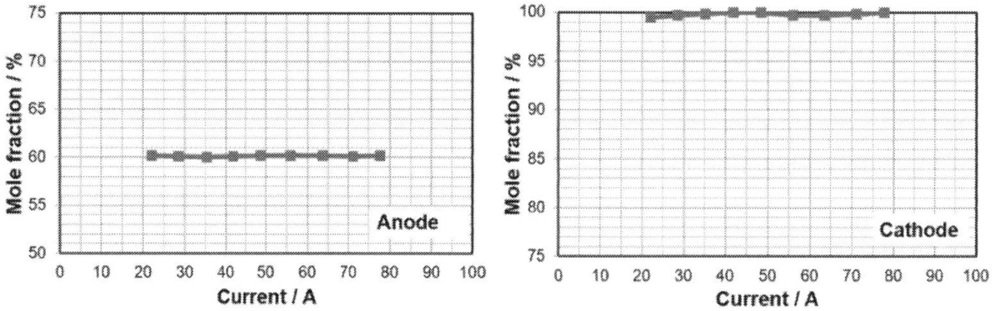

Fig.5. Mole fraction of 60 % H_2 gas mixture on the anode side and ~ 99,5 % on the cathode side as a function of the current at 80 °C and 80 % r.H.

In the simulation, the nitrogen excretion was investigated with a nitrogen content of 4 % at the anode inlet. During the pumping process, the hydrogen was pumped to the cathode side which reduced the hydrogen mole fraction at the anode along the channel. The total volume flow rate decreased and there was a nitrogen enrichment from 4 % to ~ 40 %.

Nitrogen diffusion and permeation from the anode to the cathode also happens in a HP, but in a much smaller amount than in a FC. Thus, the N_2 mole fraction is shown in Fig.6 on the anode as a percentage while the cathode side in ppm. The highest mole fraction of ~ 80 ppm of nitrogen was found in the dead end area of the cathode. At the cathode exhaust, the nitrogen content mixes with the pumped hydrogen and the concentration falls to ~ 40 ppm.

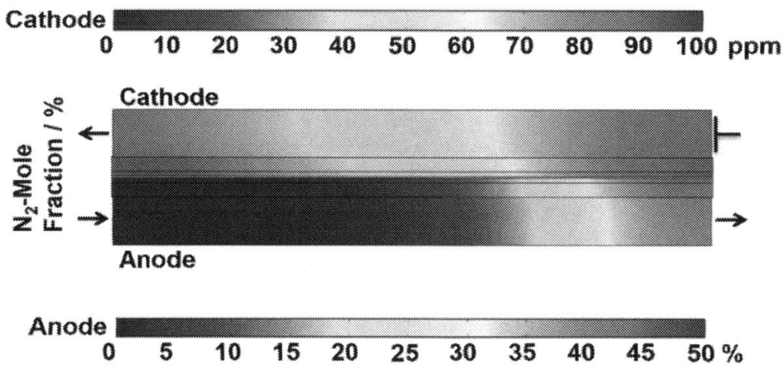

Fig.6. Results of the simulation study. Reduction of nitrogen from 4 % on the anode to ~ 40 ppm on the cathode

3.2 Water Balance

In the vehicle recirculation circuit, the N_2-contaminated hydrogen is humidified in the FC. H_2 is transported and cleaned by the HP from the anode to the cathode side. Similar to the fuel cell, in the HP water vapour is also transported through electroosmotic drag, diffusion and permeation from the anode to the cathode. The amount of water vapour transported through the HP membrane depends on the humidity supplied to the HP anode inlet. In the experiment and in the simulation, the relative humidity at the HP anode inlet was increased from 60 % to 90 % and the effects on the U-I characteristic were observed.
The aim was to achieve the highest possible relative humidity at the HP cathode output without exceeding the saturation limit. In the experiment, the best U-I characteristic was found at a relative humidity of 80 % and a temperature of 80 ° C (Fig.7).

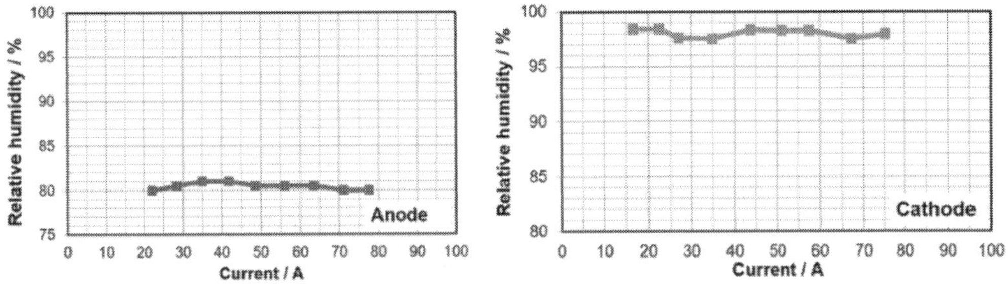

Fig.7. Results of the experimental study. The relative humidity increases from anode to cathode with the increase of the partial pressure.

In the simulation, the optimal air humidity could also be determined by the varying of the HP temperature and the relative humidity at the HP anode inlet.

The relative humidity of the HP cathode is defined by two factors. The main factor is the supply of humidified hydrogen to the HP anode inlet. The more water contained in the supplied gas, the more water is transported to the cathode side. The second factor is the increasing of the relative humidity from the anode to the cathode side due to the increase in pressure. Since the relative humidity represents the ratio of the water partial pressure to the vapour pressure, the relative humidity increases with increasing pressure.

By varying the HP temperature and relative humidity at the HP anode inlet, the optimum humidity was determined. At a relative humidity of approx. 85 % at the anode inlet with a pressure difference of 100 Pa, the gas on the cathode side did not exceed saturation. The cathode gas at the outlet has a maximum relative humidity of 92.5 % (Fig.8).

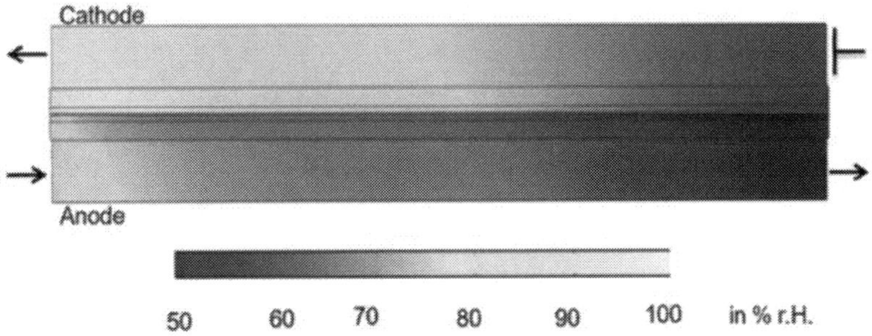

Fig.8. Results of the simulated relative humidity of the EHC.

3.3 Electric Power Consumption

Fig.9 shows the recorded U/I HP-characteristic of a 60 % H2 gas mixture at the temperature of 80 °C and a relative humidity of 80 %. It is compared with the U/I characteristic of the pure hydrogen with the same humidification and temperature. It should be noted that the flatter the U/I characteristic, the better the efficiency. Due to the lower limit of the MFCs, the range of the activating voltage at low currents is not shown.

HP studies with a hydrogen-gas mixture or pure hydrogen show increasing mass transport limitations at higher flows. Mass transport limitations due to decreasing mole fraction of hydrogen hinder the transport of the gas through the GDL to the anode. As a result, there is an increase in the voltage and energy consumption and thus, the efficiency of the Hydrogen Pump is lowered.

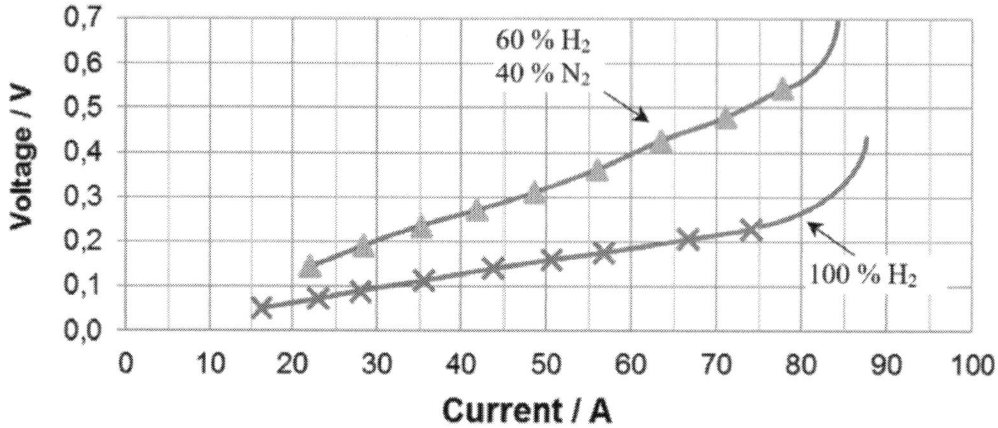

Fig.9. U / I characteristics of the HP with variation of the gas composition at 80 ° C and 80 % r.H.

To validate the use of a hydrogen pump in the vehicle system, a dynamic FC vehicle model was created consisting of the models shown in Fig.10.
This model was parameterized with real values and the dynamic behaviour as well as the fuel consumption were examined. The base vehicle for the simulation was a mid-range vehicle. The model parameters of the FC vehicle are listed in Table.1. These parameters are needed to determine the required transport processes in the FC and HP, the energy consumption and the efficiency of the powertrain in the vehicle driving cycles NEDC or WLTP.

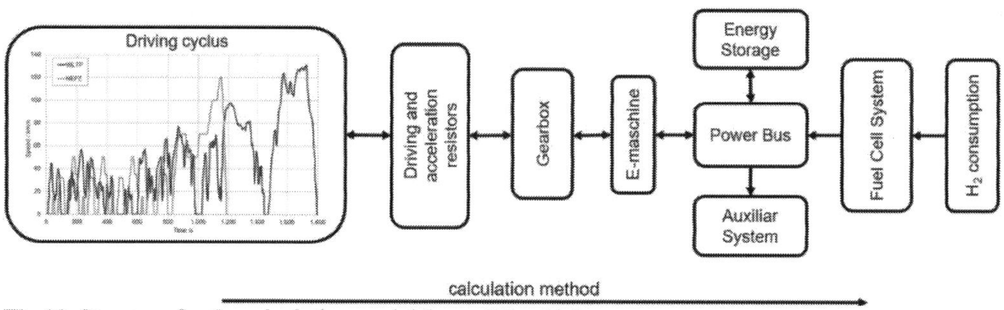

Fig.10. Structure for the calculation model for an FC vehicle

Based on the speed profile of the driving cycles, the vehicle data and the physical efficiency chain, it was possible to determine the performance losses of individual components as well as the hydrogen consumption of the entire system.
For comparison with conventional vehicles, the HP was compared to the event-driven purge strategy. In the event-driven purge strategy, the purge valve is opened when it reaches a defined nitrogen content in the anode circuit of the FC system and closed again after purging with the aim of minimizing the unused released hydrogen. In this simulation, the purge valve was opened when nitrogen content reached 5 % and closed again when it fell below 2 %.
To use the equations for the FC-vehicle simulation model, it was necessary to determine the average energy from tank to wheel in the WLTP. The main consideration of this study was the hydrogen recirculation cycle. The H2 losses purged to the environment and the energy intake of the recirculation pump were determined.

Table 1. Vehicle model parameters

Vehicle model parameters	**Symbol**	**Values**	**Units**
Vehicle total mass	m	1300	Kg
Aerodynamic drag coefficient	cw	0,25	
Frontal area	A	2	m^2
Tire drag coefficient	cr	0,012	
Tire size		215/55 R17	
Effective rolling radius of wheel		0,345	m
Auxiliaries consumption		1	kW
Walk & Slip efficiency		95 %	
Transmission efficiency		95 %	
Engine efficiency		90 %	
Converter efficiency		95 %	
Battery efficiency		90 %	

An energy flow diagram with the event-driven purge strategy with jet pump and blower is shown in Fig.11. The colour assignment represents the energy in the hydrogen in red, the electrical energy is green, and the heat or loss energy is yellow.
In this calculation, the energy consumption of the recirculation pump was 170 Wh and the purge loss was 559 Wh, which equates to 7 % of unused hydrogen from the tank.

Similarly, an energy flow diagram with the feedback with HP and jet pump is showed in Fig.12. The simulation shows that the recirculation with a HP leads to a hydrogen loss by purge of 27 Wh and the energy consumption of the HP is 83 Wh.

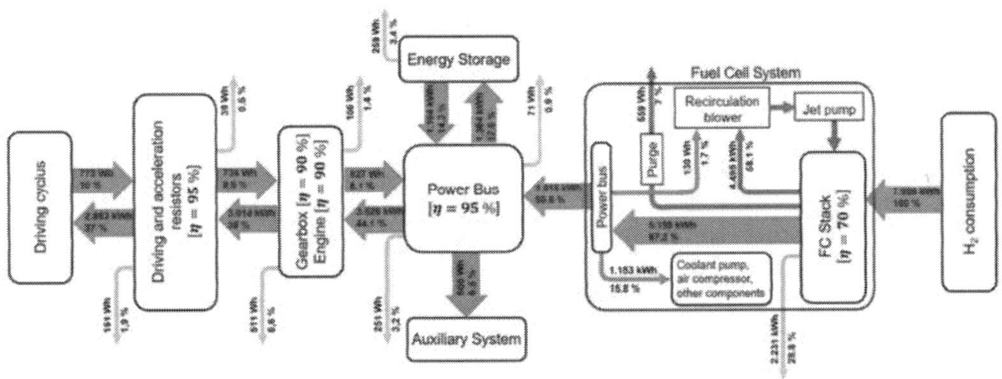

Fig.11. Energy flow on WLTP with jet pump and blower recirculation

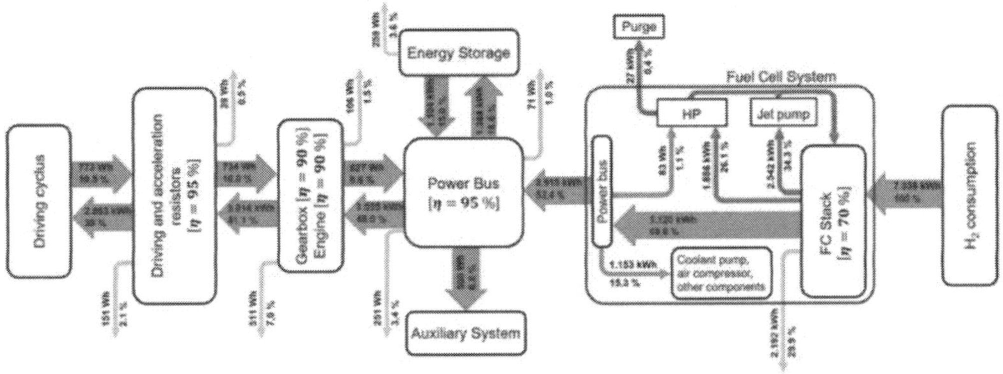

Fig.12. Energy flow on WLTP with jet pump and HP recirculation

3.4 Mechanical Integration of HP

The installation of the HP in the anode circuit of the FC system can be realized in two ways (Fig.13).
Alternative A - Here, the HP is compressed in serial with the FC stack. The result is a dual stack that is 30-60 mm longer than a typical FC stack. This structure has the best space use in the vehicle and saves additional components such as pipes and hoses. In addition, similar or same types of MEAs and bipolar plates can be used for the HP and FC.

Alternative B - Here, the HP stack is placed in front of the FC stack so that fresh hydrogen from the tank flows through the HP cathode side. On the cathode side, the pumped humidified hydrogen is mixed with the fresh hydrogen from the tank. Therefore, the dry hydrogen from the tank is humidified. As described in chapter 3.2, a water transfer takes place from the anode to the cathode side of the HP. This process is limited by the saturation of the hydrogen on the cathode side of the HP. Since dry hydrogen flows through the cathode from the tank, more water can be transported from the anode to the cathode of the HP. This leads to a better humidification of the hydrogen and to a better anode-side FC stack humidification. A reduction in size or elimination of the FC cathode humidifier can achieved with further optimization of the system.

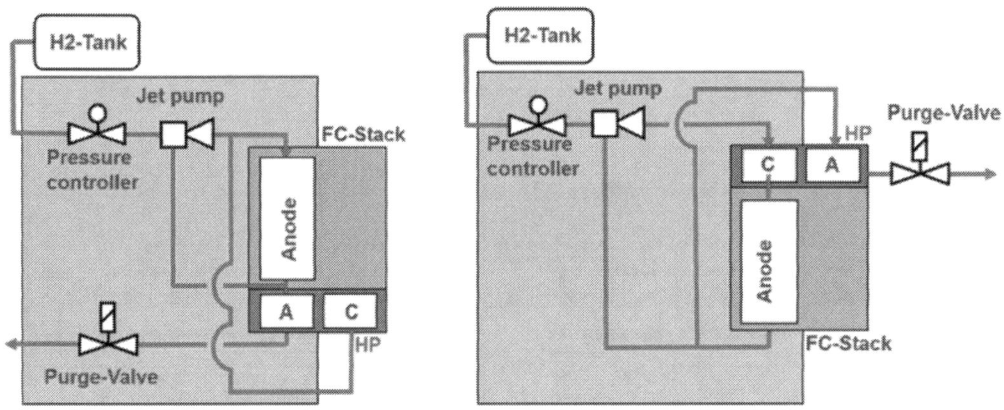

Alternative A Alternative B

Fig.13. Mechanical integration of HP in FC-system

4 Conclusions

The simulations showed that replacing the blower with the HP reduces the hydrogen loss from the purge in the WLTP from 559 Wh to 27 Wh. This means that the loss of hydrogen can be reduced by up to ~ 95 %.
Through the use of a hydrogen pump, the efficiency of the FC system could be further improved by ~2,8 % and the tank to wheel efficiency could be improved by ~2 %. For example, hydrogen consumption is lower with the HP and the efficiency of a FC vehicle could be increased from 37 % to 39 %.
Due to the capability for considerable hydrogen recovery, the HP has a significant advantage over the recirculation blower and the jet pump. The purity of the hydrogen at the cathode outlet of the HP also leads to an increased partial pressure of the hydrogen at the fuel cell and thus increases its efficiency. This results in a better overall efficiency of the FC system.
The HP has no moving parts, and thus it requires little maintenance, operates quietly, and is very robust. By using the same components as in the FC such as flow fields and GDLs, production costs can be saved. Additionally, the design in series with FC can reduce the weight and volume of the FC system, which further increase efficiencies.
Thus, the hydrogen pump has a high potential for use in the recirculation of hydrogen in FC systems in the automotive industry and is a great alternative to today's recirculation blower. The HP could play a vital role in the evolution of fuel cell vehicles and help to decrease the hydrogen consumption, resulting in more energy-efficient and greener mobility in the future.

References

[1] F. Barbir, H. Görgünm, *Electrochim. Acta* **2007**, *37,* 359.
[2] J.M. Sedlak, J.F. Austin, A.B. LaConti, *J. Hydrogen Energy* **1981**, *6,* 45.
[3] B. Rohland, K. Eberle, R. Ströbel, J. Scholta, J. Garche, *Electrochim Acta,* **1998**, *43,* 3629.
[4] R. Ströbel, M. Oscipok, M. Fasil, B. Rohland, L. Jörissen, J. Garche, *J. Power Sources* **2002**, *105,* 208.
[5] T. Schwarz, *Ph. D. Thesis,* Technische Universität, Clausthal, Germany, **2010**.
[6] R. Staeck, *Ph. D. Thesis*, Technische Universität, Braunschweig, Germany. **2013**.
[7] O. Berger, *Ph. D. Thesis,* Universität, Duisburg-Essen, Germany, **2009**.
[8] M. Mayur, S. Strahl, A. Husar, W. Bessler, *J.Hydrogen Energy,* **2015**, *40,* 16466.

Fabrication of alloy foam-supported solid oxide electrolysis cell (SOEC) for hydrogen production

Ramin Visvanichkul[1], *Pramote* Puengjinda[2], *Tara* Jiwanuruk[2], *Saranya* Peng-Ont[2], *Nichaporn* Sirimungkalakul[2], *Watcharin* Ngampuengpis[2], *Thana* Sornchamni[2], *Pattaraporn* Kim-Lohsoontorn[1,*]

[1]Centre of Excellence on Catalysis and Catalytic Reaction Engineering, Department of Chemical Engineering, Faculty of Engineering, Chulalongkorn University, Bangkok, 10330, Thailand
[2]PTT Innovation Institute, PTT Public Company Limited, 13170, Thailand

Abstract. Alloy foam-supported SOEC is fabricated. Nickel-iron (Ni-Fe) alloy foam (Porosity: 5-130 ppi) is used for cell support. Single thin-cell composed of Ni- $Sc_{0.1}Ce_{0.05}Gd_{0.05}Zr_{0.89}O_2$ (SCGZ) cathode, SCGZ electrolyte and $Ba_{0.5}Sr_{0.5}Co_{0.8}Fe_{0.2}O_{3-\delta}$ (BSCF) anode is fabricated. Electrode powders are mixed with additives forming as slurry for wet chemical coating. 70%weight content of cermet provides smooth surface and sufficient viscosity to prevent slurry sweep through the porous foam. However, severe cracking is clearly seen on the surface of the cell because of mismatching of thermal expansion coefficient (TEC) during sintering. Therefore, the cell with three cathode layers having TEC gradient (13.83, 13.62 and 13.40 $ppmK^{-1}$) and %weight content of cermet gradient (70%, 60% and 50%weight) is fabricated. Heating rate and steps are controlled at 0.5°C/min (600 °C), 3°C/min (800 °C) and 1°C/min (1,300°C, 4 h) to burn off additives before sintering.

1 Introduction

Hydrogen is a promising a carrier for clean and sustainable energy. It is also used as a feedstock for various chemical production. The demand tends to increase continuously. Steam electrolysis through solid oxide electrolysis cell (SOEC) can produce high purity hydrogen. Challenge remains for SOEC development is to improve performance and durability of the cell. There are various strategies for SOEC development such as material selection, sealant development, and optimized operating condition. The reduction of cell's resistance by making cell thinner is one among strategies. In general, the cell requires a support from a thick electrode and is called as cathode- or anode-supported cell [1-3]. Therefore, to fabricate a thin cell, metal support is extensively studied to reduce ohmic resistance in thick electrode and increase strength in cell. Nevertheless, metal support SOEC is still facing high fabrication cost and poor durability in long-term operation due to metal oxidation under high steam containing environment. Using an alloy as a support to

[*] Corresponding author: pattaraporn.k@chula.ac.th

fabricate a thin cell can provide lower resistance and oxidation rate and help enhance mechanical properties of cell [4-7].

In this study, alloy-foam supported SOEC was fabricated. Nickel-iron (Ni-Fe) alloy foam (Porosity: 5-130 ppi) was used for the cell support. Single thin-cell composed of Ni-$Sc_{0.1}Ce_{0.05}Gd_{0.05}Zr_{0.89}$ (SCGZ) cathode, SCGZ electrolyte and $Ba_{0.5}Sr_{0.5}Co_{0.8}Fe_{0.2}O_{3-\delta}$(BSCF) anode was fabricated. The morphology of the cell was characterized by scanning electron microscopy (SEM). The electrochemical performance of the SOEC is measured under H_2O/H_2 feed with varied steam content in a controlled temperature from 600°C to 900 °C.

2 Experimental

2.1 Substrate Preparation

NiFe foam (Tmax Battery Equipments Limited, China) having porosity of 5-130 ppi and 3 mm thickness was used as substrate. The foam was cut into 25 mm diameter with a thickness of 3 mm. After that, the foam was uniaxially pressed at 62.5 psi and for 30 s to produce smoother surface, ready for electrodes layer deposition.

2.2 Slurry Preparation

2.2.1 Cathode Slurries

NiO (Sigma-Aldrich, USA) and $Sc_{0.1}Ce_{0.05}Gd_{0.05}Zr_{0.89}O_2$ (SCGZ, Kceracell, Republic of Korea) were used as cermet starting powders for fabrication of cathode slurry. The powders were mixed using ball-milling in solvent consisting of toluene and ethanol at room temperature for 24 h. Then, plasticizer and dispersant were added into the mixture and the compositions were mixed for another 24 h. Polyethylene glycol 400 (PEG 400) and α-terpineol were used as plasticizer and dispersant, respectively. Cermet content in slurry was varied between 40-80 wt%. NiO content in cermet was 66-75 wt%.

2.2.2 Electrolyte Slurries

Electrolyte slurry was obtained by well mixing of SCGZ powder, solvents and additives at 70°C for 24 h. Xylene and Butyraldehyde were used as solvent. Additives were included Polyvinyl butyral (B-98, Sigma-Aldrich), PEG 400 and Polyvinylpyrrolidone (PVP, Sigma-Aldrich) as binder, plasticizer and dispersant, respectively. 0.5 wt% CuO (Sigma-Aldrich) was added into the electrolyte slurry as sintering additive to help decrease sintering temperature. The effect of CuO sintering additive on sinterability and electrochemical performance of the cell was reported in our previous work [8].

2.2.3 Anode Slurries

$Ba_{0.5}Sr_{0.5}Co_{0.8}Fe_{0.2}O_{3-\delta}$ (BSCF, Kceracel, Republic of Korea) were used as anode material. The anode powder was mixed with α- terpineol in weight ratio 1:1 to form anode paste.

2.3 Fabrication of NiFe-alloy-foam-supported SOEC

Cathode slurries were deposited onto foam substrate by tape-casting technique. The green tape were dried at room temperature and fired at 1100°C for 2 h. Then, SCGZ thin layer was coated by electrolyte slurry dip-coating, followed by sintering at 1250-1300°C for 4 h. Heating rate and steps which were 3 condition were investigated at 1) 2°C/min (300°C, 5 h) and 3°C/min (1250°C, 4 h), 2) 1°C/min (300°C, 5 h) and 2°C/min (1250°C, 4 h), 3) 0.5°C/min (600°C), 3°C/min (800°C) and 1°C/min (1250°C 4 h). Anode slurry was screen printed on the sintered cell. The schematic drawing of the cell is shown in Fig 1. Microstructural images of sintered cell was taken by using SEM (Hitachi S-3400N JEOL model S-3400, Japan).

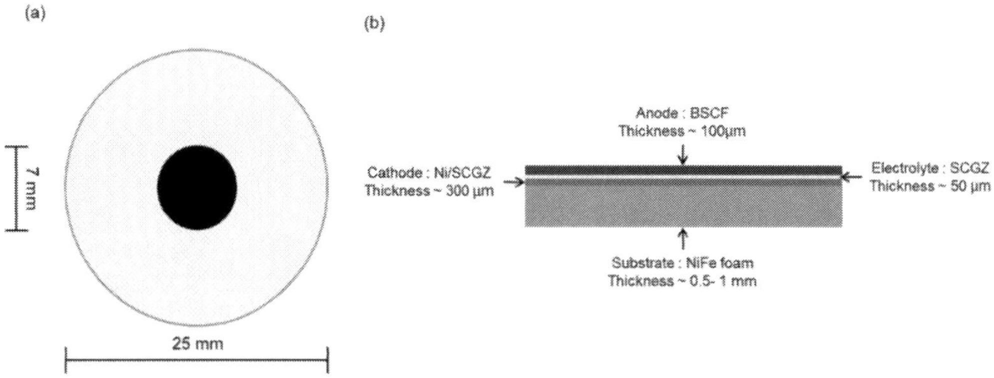

Fig. 1 Schematic drawing of Alloy foam-supported solid oxide electrolysis cell (SOEC); (a) top view, (b) side view.

2.4 Electrochemical Performance Measurement

The fabricated cell was installed in test station consisting of cell holder with high temperature sealant (Ceramabond 552, Aramco) inside a furnace (Chavachote). Platinum (Pt) conductive paste (70 wt% Pt, Nexceris), mesh and wire (Kceracell) were attached on the fabricated cell for electrical connection. The electrochemical performance was measured at 600-900°C with feed containing 60-80 vol% of steam and constant hydrogen content (balanced N_2). Deionized water was supplied for steam generation by high performance liquid chromatography pump (Teledyne SSI, USA). A heated-tube was used as steam generator. Linear sweep current techniques was applied to generate current/voltage (I/V) curves. A potentiostat (PG310 Metrohm Autolab, Netherlands) was used to control the potential from 0.4 V to 1.8 V with a scan rate of 20 mVs^{-1}.

3 Results and Discussion

3.1 Cermet content in cathode slurry

Unfired cathode with 60 wt% Ni / 40 wt% SCGZ on NiFe-alloy foam with varied cermet content in cathode slurry are shown in Fig. 2. After casting the cathode, it was found that 70 wt% cermet in slurry provided smooth surface and sufficient viscosity to prevent slurry sweeping through the porous foam.

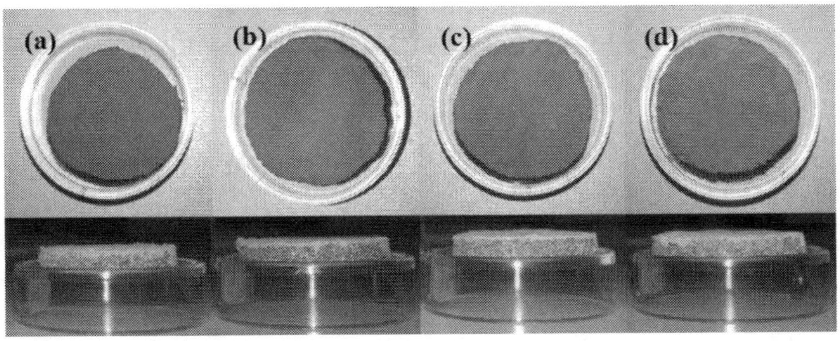

Fig. 2 Unfired Ni-SCGZ cathode on NiFe-alloy foam with vary cermet content in cathode slurry: (a) 40%wt, (b) 50%wt, (c) 70%wt, and (d) 80%wt cermet content

Electrolyte was coated on fired cathode and sintered at 1300°C. Sintered half cell with 70 wt% cermet content in cathode slurry and the SEM image are shown in Fig. 3. Fig. 3 reveal that thin SCGZ electrolyte was deposited on cathode layer. However, severe cracking was clearly found all over the surface of the cell. This was likely due to mismatching of thermal expansion coefficient (TEC) among electrolyte (10.65 ppm/K), cathode (13.40 ppm/K), and substrate (14.62 ppm/K). Gondolini et al.[6] investigated Ni-GDC layer deposition on NiCrAl foam for SOFC application. It was reported that sintered layer which is accompanied by a certain shrinkage lead to crack formation due to TEC mismatching. To overcome this problem, cathode was divided into three layers with vary cermet content in slurry (70 wt%, 60 wt% and 50 wt%) and NiO content in cermet content (75 wt%, 70 wt% and 66 wt%). Three slurries were formulated for fabrication of cathode layers as presented in Table 1.

Table 1. 1 Cathode slurries properties

Layers / **Properties**	**NiO- SCGZ 1**	**NiO- SCGZ 2**	**NiO- SCGZ 3**
TEC (ppm/K)	13.83	13.62	13.40
Viscosity (Pa·s) at shear rate 10 s^{-1}	Undefined	702.1	578.3
Cermet content in slurry	70%wt	60%wt	50%wt

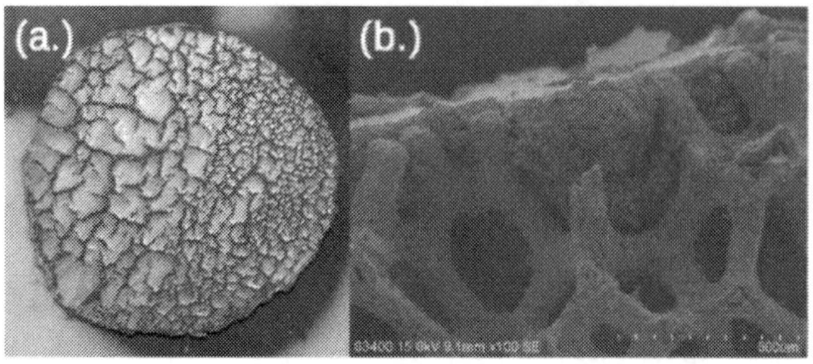

Fig. 3 Sintered half cell of Ni-SCGZ/SCGZ on NiFe-alloy foam: (a) Surface image and (b) SEM image

3.2 Heat treatment in sintering process

Heating treatment in electrolyte sintering process is one of the important factors for cell fabrication. The effect of heating rate during sintering was compared. Sintered half cells in different heating rate are shown in Fig. 4.

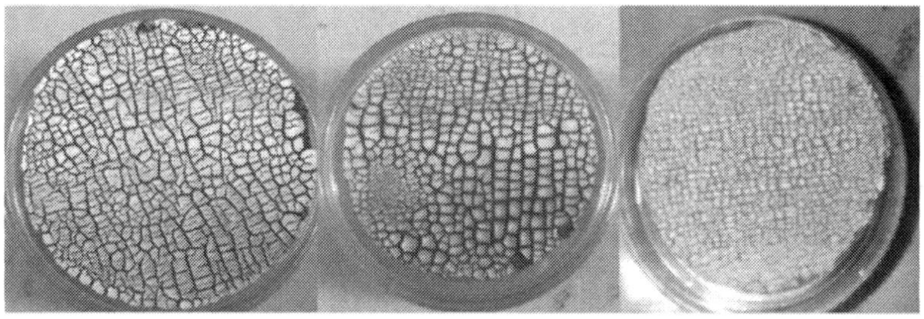

Fig. 4 Sintered half cell of Ni-SCGZ/SCGZ on NiFe-alloy foam using different heating rate: a) 2° C/min (300°C 5 h) and 3° C/min (1250 °C 4 h), (b) 1° C/min (300 °C 5 h) and 2° C/min (1250 °C 4 h), (c) 0.5° C/min (600 °C), 3° C/min (800 °C) and 1° C/min (1250 °C 4 h)

Fig. 4 reveals that fast heating rate dominated the crack on the cell surface, corresponding to the work of Jin et al.[9] who studied the fabrication of Ni-SSZ porous anode for anode-supported SOFC fabricated by tape casting and co-firing technique. It was proposed that low heating rate could suppress the defect formation and delamination. The mismatch in shrinkage of anode and electrolyte were accelerated at faster heating. In this study, slow heating and steps to burn off additives were applied in heat treatment.

3.3 Microstructure of alloy-foam supported SOEC

The NiFe-alloy-foam/Ni-SCGZ/SCGZ/BSCF was successful fabricated by adjusting TEC in cathode, using suitable heat treatment during sintering process, and preparing foam substrate. Cathode was divided into three layers with different composition to form different TEC. Fig. 5 shows the cell during preparation and Fig. 6 presents the cell's morphology. SEM images revealed well attachment between all layer (electrolyte, electrodes and substrate). However, as can be seen in Fig. 6 (a), some porosity on electrolyte layer was observed due to insufficient electrolyte densification.

 https://doi.org/10.1051/e3sconf/202015501002

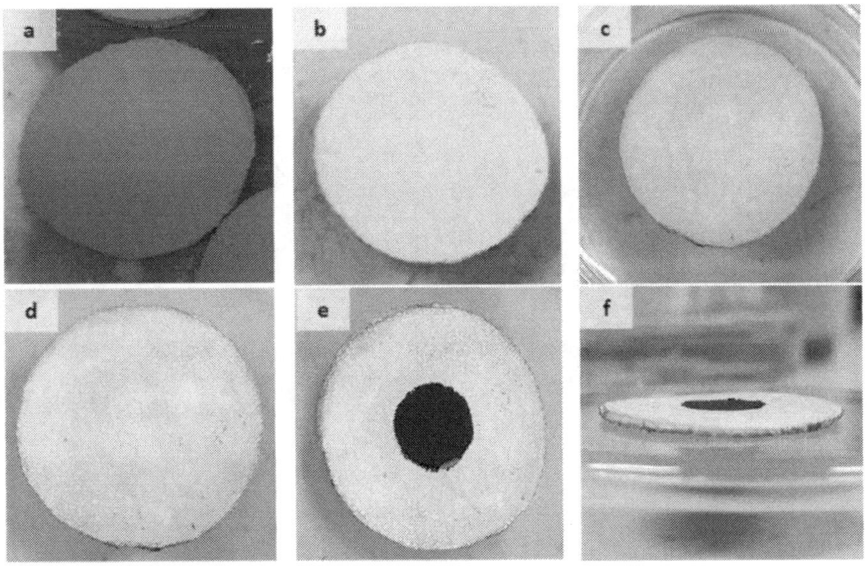

Fig. 5 NiFe alloy foam/Ni-SCGZ/SCGZ/BSCF SOEC: (a) Cathode layer deposition, (b) sintered cathode, (c) electrolyte deposition, (d) sintered electrolyte, (e) anode deposition, and (f) side view of the cell after heat treament

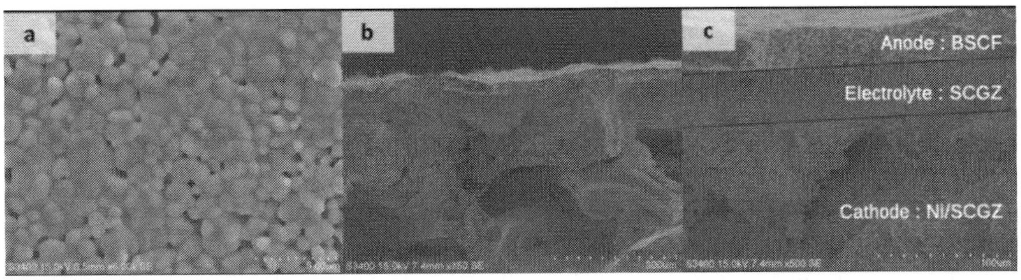

Fig. 6 SEM images of compressed NiFe-foam supported solid oxide electrolysis cell: a) SCGZ electrolyte sintered at 1,300 °C, (b) and (c) Anode/Electrolyte/Cathode/Support layers

3.4 Electrochemical Performance measurement

The electrochemical performance of the alloy-foam supported SOEC was measured at 800°C with 80% steam content using linear sweep current techniques (Fig. 7). Open circuit voltage (OCV) of cell (~0.5V) was dropped below theoretical value. Lenser et al.[10] were reviewed that OCV is either decreased by the external load powered by the cell (SOFC) or increased by an external power source (SOEC) and loss of concentration of reactant. In this study, low OCV was likely due to insufficiently densified electrolyte which was sintered at 1,300 °C. Ohmic resistance of cell in SOEC mode was also higher than SOFC mode. The current density at 1.1 V which is operating voltage in SOEC was consumed 0.3 A/cm^2.

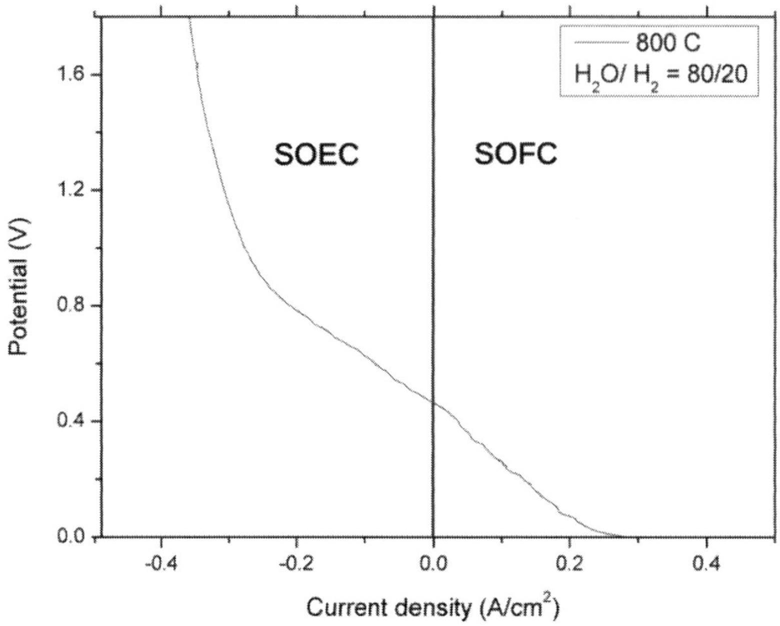

Fig. 7 I-V Curve in 80%H_2O and 20%H_2 for NiFe foam/ Ni-SCGZ/ SCGZ/ BSCF

4. Conclusion

NiFe-alloy-foam supported Ni-SCGZ/SCGZ/BSCF solid oxide electrolysis cell was fabricated by wet chemical coating of electrodes slurry. There were various parameters affected the fabrication process such as heat treatment and slurry composition. The TEC mismatching among cathode, electrolyte, and substrate was found, leading to a crack in the cell after heat treatment. Therefore, cathode layers were designed to obtain TEC gradient to suppress TEC mismatching in the cell. Three layers of cathode with different compositions were formed. Moreover, fast heating rate during sintering was found to dominate the shrinkage behaviour and led to the cell defect. Slow heating rate with a step to burn off the additives could decrease the defect in sintered cell. However, microstructure of fabricated cell revealed some porous in electrolyte layer and likely to be the cause of OCV dropping during electrochemical performance measurement. Sintering temperature at 1,300°C for electrolyte slurry having 0.5%wt CuO as sintering additive was insufficient for the electrolyte densification.

References

1. E. Ioannidou, C. Neofytidis, L. Sygellou, D.K. Niakolas, Applied Catalysis B: Environmental, **236**, 253-64, (2018).
2. D. Dong, X. Shao, X. Hu, K. Chen, K. Xie, L. Yu, Z. Ye, P. Yang, G. Parkinson, C.-Z. Li, Int J Hydrogen Energy, **41**, 19829-35, (2016).
3. A. Mahmood, S. Bano, J.H. Yu, K.-H. Lee, Journal of CO2 Utilization, **33**, 114-20, (2019).

4. S. Molin, B. Kusz, M. Gazda, P. Jasinski, J Power Sources, **181**, 31-7, (2008).
5. K. Li, X. Wang, L. Jia, D. Yan, J. Pu, B. Chi, L. Jian, Int J Hydrogen Energy, **39**, 19747-52, (2014).
6. A. Gondolini, E. Mercadelli, A. Sangiorgi, A. Sanson, J Eur Ceram Soc, **37**, 1023-30, (2017).
7. N. Xu, M. Chen, M. Han, J Alloys Compd, **765**, 757-63, (2018).
8. R. Visvanichkul, P. Puengjinda, T. Jiwanuruk, S. Peng-Ont, N. Sirimungkalakul, W. Ngampuengpis, h. Sornchamni, P. Kim-Lohsoontorn, Processes, **7**, 868, (2019).
9. C. Jin, Y. Mao, N. Zhang, K. Sun, Int J Hydrogen Energy, **40**, 8433-41, (2015).
10. C. Lenser, D. Udomsilp, N.H. Menzler, P. Holtappels, T. Fujisaki, L. Kwati, H. Matsumoto, A.G. Sabato, F. Smeacetto, A. Chrysanthou, S. Molin, *Advanced Ceramics for Energy Conversion and Storage*, 387-547, (2020).

Study on Hybrid Energy Storage Configuration and Control Strategy of Grid-connected Wind Hydrogen System

Xiaojun Shen [1,*], *Congying* Nie [1], *Hong* Lv [2], *Wei* Wang [3]

[1]Department of Electrical Engineering, Tongji University, Shanghai 201804, China
[2]Automotive College, Tongji University, Shanghai 201804, China
[3]State Grid Corporation, Beijing 100031, China

Abstract. The randomness and volatility of wind energy bring great challenges to wind power grid-connected. The hybrid energy storage technology based on electrolysis cell hydrogen production and super capacitor has become an effective way to stabilize wind power fluctuations. Based on the establishment of the grid-type wind-hydrogen coupling system working characteristic constraint and the minimum initial investment cost of the hybrid energy storage system, an alkaline electrolytic cell-super capacitor hybrid storage based on low-pass filtering-fluctuation observation is proposed. The method can be configured, and a hybrid energy storage coordinated control strategy based on the super capacitor SOC (state of charge) is developed. The case study results show that the hybrid energy storage system configuration method and control strategy proposed in this paper are effective, which can reduce the fluctuation of wind power grid-connected power to meet the grid connection standard.

1 Introduction

Wind power is the main form of utilization of renewable energy. Its rapid and large-scale development makes it difficult for grids to absorb wind power and the problem of abandoning wind power is getting worse. Hydrogen energy is characterized by clean, pollution-free, highly efficient and renewable. It is regarded as the most potential energy carrier in the future. The use of electrical energy generated by renewable energy to electrolyze water to produce hydrogen is considered to be the best way to lead to hydrogen economy [1]. The related theoretical research and engineering demonstration application results show that the combination of electrolyzed water hydrogen production and wind power can effectively stabilize the wind power volatility and enhance the schedulability of

* Corresponding author: xjshen79@163.com

wind power, and provide a new technical solution for solving wind power consumption problems, which is a research hotspot at home and abroad [2].

The literature search results show that since the beginning of this century, wind power hydrogen production technology has gradually gained worldwide attention, and related research has been carried out. The early wind hydrogen system demonstration project focused on wind power, hydrogen production (electrolysis tank), hydrogen storage and direct simple power control [3]. In recent years, with the rapid development of wind-hydrogen system demonstration projects, research has become more and more in-depth, and rich research results have been achieved in hydrogen production power conversion technology, system coordination control, capacity allocation, program planning and technical economic analysis. For example, the reference [4] proposes a smooth control strategy for multi-type energy storage systems using real-time wavelet transform for frequency allocation and double-layer fuzzy control theory for energy storage power correction. In reference [5], the spectrum analysis method is used to optimize the energy storage capacity distribution of the energy storage system, which can effectively reduce the fluctuation of wind power and ensure that the energy storage equipment will not be over flushed or over discharged. In reference [6], the mathematical model of the hybrid system of wind power / photovoltaic / hydrogen production / super capacitor is established, and the corresponding control strategy is proposed for different operation conditions. The operation efficiency of the system is improved, the DC bus voltage is stable, and the operation reliability of the system is guaranteed; The energy storage forms used in reference [7] are electrolytic cell, hydrogen storage tank and proton exchange membrane fuel cell cycle system, which are connected to the DC side of the power converter of the wind power system to form the wind power hydrogen cogeneration system, and the coordinated control strategy of each component unit is designed. The simulation results show that the proposed control strategy is effective; In reference [8], based on the DC coupled wind, hydrogen and super capacitor hybrid power system, two power management strategies are proposed and compared through experiments. It is found that the "source tracking" strategy has better performance than the "grid connected" strategy in power grid regulation. In reference [9], based on the analysis of the spectrum characteristics of wind power, the wavelet packet decomposition technology is used to decompose it into low-frequency grid connected part and high-frequency part, and then the probability density function of fluctuating power is fitted by Fourier approximation to obtain the expected energy storage capacity configuration and optimize the energy storage capacity. In reference [10], taking the maximum profit of the system as the objective, considering environmental benefits, government subsidies and time value of funds, the objective function and its constraints are established, and the economic evaluation is carried out by using the improved chemical reaction algorithm. However, the capacity of hydrogen production and fuel cell cannot completely smooth the fluctuation of wind power in the maximum profit, and other energy storage equipment needs to be added.

The results of literature analysis show that the existing research results consider the characteristics of electrolytic cell less, which is not conducive to engineering application. In view of this, based on the study of the wind hydrogen coupling system and the actual external characteristics of the alkaline electrolyzer, the configuration method and process of the electrolyzer super capacitor hybrid energy storage system are established based on the low-pass filter volatility observation method, and the optimal coordination control strategy is formulated. Finally, the effectiveness of the configuration method and control strategy is verified by the case simulation.

2 Topology and constraints of grid connected wind hydrogen coupling system

2.1 Overview of grid connected wind hydrogen coupling system

Wind hydrogen coupling power generation has become an important means to solve the "bottleneck" problem of wind power connected to grid. As a kind of clean energy, hydrogen can feed back electric energy to the power grid through hydrogen fuel cell in the low period of power generation, or enter the industrial and commercial fields through vehicle or pipeline. At the same time, the wind hydrogen coupling system will greatly promote the development of hydrogen fuel cell vehicle industry rapidly [11]. The typical wind hydrogen coupling hybrid energy storage system includes wind turbine, electrolytic water hydrogen generation unit and super capacitor energy storage unit, as shown in Figure 1.

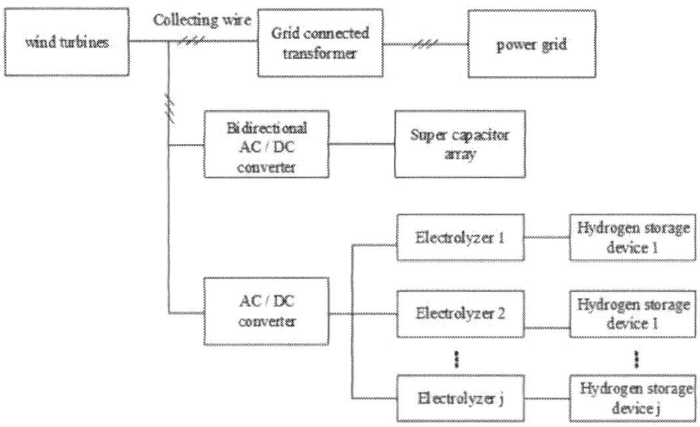

Fig.1. The structural diagram of the wind-hydrogen coupling system

The electrolytic cell and super capacitor are connected with the wind power system through AC / DC device, in which the electrolytic cell is the main energy storage equipment as the cost of it is low, but due to the lack of absorption for high-frequency power, super capacitor is introduced as the auxiliary energy storage equipment [12].

2.2 working characteristics constraints of wind hydrogen coupling system

2.2.1 working characteristics constraints of electrolyzer

According to the relevant research results of the external characteristics of the alkaline electrolyzer, the external characteristics of the electrolyzer can be expressed as [13]:

$$\begin{cases} V \in [V_{\min}, V_{\max}] \\ T \in [T_{\min}, T_{\max}] \\ I \in [I_{\min}, I_{\max}] \\ P \in [P_{\min}, P_{\max}] \\ \Delta t \le 1ms \end{cases} \tag{1}$$

Where, V is the working voltage; I is the working current, and the range corresponds to the voltage; T is the working temperature; Δt is the response time; P is the working power, and its lower limit value is not lower than the thermal insulation power P_{min}. Thermal insulation power is defined as the balance of heat production and heat dissipation of the electrolyzer under this power condition, and the temperature stability of the electrolyzer can be guaranteed without external environmental control device. Only when the power of the cell is not lower than this value can the cell have a millisecond level response speed.

2.2.2 working characteristics constraints of super capacitor

As a kind of energy storage equipment with excellent properties, super capacitor has the characteristics of fast response and high efficiency. However, the biggest disadvantage of limiting its application is its high price, and its energy unit price is about 30 times of the traditional lithium-ion battery [14]. In consideration of the economy, the super capacitor, as an auxiliary energy storage device, generally occupies a small part of the capacity in the hybrid energy storage system. In general, super capacitor exists in the form of modularization, through different number of modules in series or parallel to get the corresponding demand energy storage system. In the engineering application of super capacitor, the constraint of formula (2) should be satisfied.

$$\begin{cases} E < E_{\max} \\ U < U_N \\ I < I_{\max} \\ SOC \in [SOC_{\min}, SOC_{\max}] \end{cases} \tag{2}$$

Where E_{max} is the maximum value of stored energy; U_N is the rated voltage; I_{max}

is the maximum value of allowable passing current; SOC is the state of charge of super capacitor. In order to ensure the normal and efficient operation of super capacitors and good service life, SOC of general super capacitors has a certain range of constraints.

2.2.3 characteristics constraints of wind turbine and grid connection

Without considering the wake and electrical losses of the wind farm, the output power of the wind farm is equal to the sum of the output of all the fans in the field, and the power output of the fans mainly depends on the wind speed at the height of the fan hub. The relationship between the two can be expressed by the following subsection function [15]:

$$P_w = \begin{cases} 0 & V<V_{in}\text{或 } V_{out}<V \\ \dfrac{P_e(V^3-V_{in}^3)}{V_{out}^3-V_{in}^3} & V_{in}<V<V_e \\ P_e & V>V_e \end{cases} \tag{3}$$

Where V is the wind speed at the wind turbine; V_{in} is the cut-in wind speed; V_{out} is the cut-out wind speed; V_e is the rated wind speed; P_e is the rated output power; P_w is the actual output power of the wind turbine.

In order to ensure the safety and reliability of wind power connected to grid, all countries have developed corresponding technical indicators for wind power connected to grid. China has also formulated relevant regulations for the grid connection of wind farms with different installed capacity, and the maximum change of output active power of wind farms must be within the maximum change limit of 1min and 10min active power [16].

$$|P_{wind}-P_{el}-P_{sc}|\le P_{\max} \tag{4}$$

In the formula, P_{wind} is the original power of wind farm; P_{el} is the absorption power of electrolytic cell; P_{sc} is the absorption power of super capacitor; P_{max} is the maximum limit value of power fluctuation specified in national standard, and there are two different values on the time scale of 1min and 10min.

Table 1. Maximum limit of wind farm active power change

Installed capacity of wind farm (MW)	10min active power Change value (MW)	1min active power Change value (MW)
<30	10	3
30~150	Installed capacity/3	Installed capacity/10
>150	50	15

2.3 Minimum initial investment cost constraint

In this paper, the main goal is to stabilize the power fluctuation of wind power by electrolyzer and super capacitor. Under this condition, the economy of the hybrid energy storage equipment is considered. Therefore, the profit of hydrogen production is not considered in the evaluation of the economy index, and the initial investment cost of the system is minimized to optimize the configuration scheme.

Investment cost of hydrogen production equipment [17]:

$$F_{EL} = P_{EL} u \tag{5}$$

In the formula, u is the construction cost of hydrogen production equipment per unit capacity; P_{EL} is the power of the electrolytic cell.

The investment cost of super capacitor can be determined according to the capacity [14]:

$$F_{SC} = W_{SC} f_n \tag{6}$$

In the formula, W_{SC} is the capacity of the super capacitor; f_n is the capacity unit price of the super capacitor.

According to the market of super capacitor and electrolyzer, the cost of super capacitor is much higher than that of electrolyzer when the same amount of electricity is stored. Therefore, in order to minimize the cost of the hybrid energy storage system, the electrolyzer is preferred as the main energy storage equipment and the super capacitor as the auxiliary energy storage equipment. At the same time, because of the particularity of the electrolyzer, the negative value of the power fluctuation, that is to say, the energy storage system needs to provide power to the grid, which must be borne by the super capacitor, which requires that the capacity of the super capacitor cannot be lower than a certain limit value. In order to make the system meet the grid connection standard and consider the economy, it is necessary to find out the minimum limit value of super capacity and set its capacity as this value.

$$\begin{cases} W_{SC} = W_{SC\min} \\ F_{\min} = F_{EL\max} + F_{SC\min} \end{cases} \tag{7}$$

Where, W_{SCmin} represents the minimum capacity of the super capacitor that meets the conditions; F_{min} represents the minimum total cost of the initial investment of the system; F_{ELmax} represents the maximum cost of the corresponding electrolyzer, and F_{SCmin} represents the minimum cost of the super capacitor.

3 Configuration of hybrid energy storage system based on low pass filter and fluctuation observation

The configuration roadmap of the wind power fluctuating energy storage system adopted in this paper is shown in Figure 2. First, the original wind power to be suppressed is divided into the allowable fluctuating part and the exceeding part according to the national grid connection standard by algorithm; then, the exceeding part is further divided into low frequency component and high frequency component by spectrum analysis, and the low frequency component focuses on the energy density and high frequency component focuses on the power density of the energy storage medium. Finally, according to the characteristics of the energy storage medium, the matching selection is carried out to complete the configuration of the mixed energy storage.

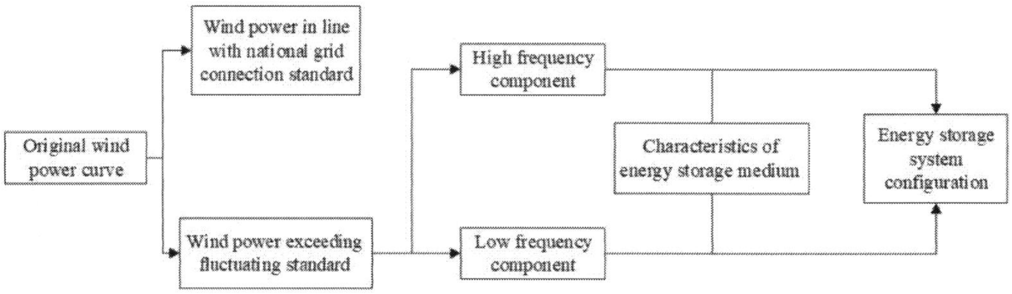

Fig.2. Configuration Technology Roadmap

3.1 Low pass filtering

In this paper, the first-order low-pass filtering method is used to analyze the power fluctuation of wind power to be suppressed. The fluctuating power is regarded as a fluctuating signal, so the low-pass filtering is used to filter out the high-frequency components, so as to reduce the power fluctuation. The traditional low-pass filtering methods are as follows [18]:

The first-order low-pass filter method is mainly improved from the first-order low-pass filter, and its mathematical expression is as follows:

$$\tau * \frac{dU_o}{dt} + U_o = U_i \tag{8}$$

Where $\tau = RC = 1/2\pi f_c$ is the time constant of the low-pass filter, and f_c is the cut-off frequency of the filter. U_o is the output signal and U_i is the input signal. When applied to the fluctuation of wind power, the input signal is the original fluctuation power P_w, and the output signal is the power P_o connected to the grid after being flattened. The purpose of this paper is to make the fluctuation of grid connected power P_o smaller and meet the corresponding grid connected standards.

Let the period of wind power sampling be T_s, and discretize formula (8) to obtain:

$$\tau \frac{P_o(t) - P_o(t - T_s)}{T_s} + P_o(t) = P_w(t) \tag{9}$$

From equation (9):

$$P_o(t)=\frac{\tau}{\tau+T_s}P_o(t-T_s)+\frac{T_s}{\tau+T_s}P_w(t) \tag{10}$$

Since the change range of time constant τ is $[0,\infty)$, it is difficult to optimize the adjustment, so that let $\alpha = T_s/\tau+T_s$, α is called the filter coefficient, then equation (10) can be written as:

$$P_o(t)=(1-\alpha)P_o(t-T_s)+\alpha P_w(t) \tag{11}$$

It can be seen from equation (11) that the larger the filter coefficient is, the greater the impact of input power on output power is; the smaller the filter coefficient is, the greater the impact of historical value of output power on output power is, the smoother the curve of output power is, and the better the filtering effect is.

3.2 Window sliding wave observation method

In order to analyze the fluctuation of power, judge whether it meets the national grid connection standard, and find the best filter coefficient to minimize the initial investment cost of the system. In this paper, a window sliding observation method is proposed to extract data from the power waveform and analyze the data. The specific methods are as follows: firstly, continuous sampling data at 1 min interval and analysis it. According to the definition of 10min wave momentum, it is necessary to collect the power data in the first ten minutes of each time, find the maximum and minimum values of these ten data, take the difference as the maximum value of 10min wave momentum at this time and store it in the array. The width of these 10 data is the width D of the established window. As time goes on, the window also moves together, so that 10 min fluctuation of each time can be obtained.

Set t_1, t_2, $t_3 \cdots t_n$ is the sampling time, where $n \in [1,600]$. The corresponding sampling values are S_{t_1}, S_{t_2}, $S_{t_3} \cdots S_{t_n}$. For any time $t_n(n > 10)$, the fluctuation in 10 minutes is as follows:

$$A_{t_n}=Max(S_{t_{n-10}}\cdots S_{t_n})-Min(S_{t_{n-10}}\cdots S_{t_n}) \tag{12}$$

Among them, $\mathrm{Max}(S_{t_{n-10}}\cdots S_{t_n})$ represents the maximum value of data in this time window, $\mathrm{Min}(S_{t_{n-10}}\cdots S_{t_n})$represents the minimum value of data in this time window.

In the same way, for 1 min fluctuation, just adjust the window size D to 2. Figure 3 is the schematic diagram of the window sliding wave observation method. The red box represents the window size set up. As time goes by, the window moves with it, so that only the data in the window needs to be analyzed to get the required wave fluctuation information.

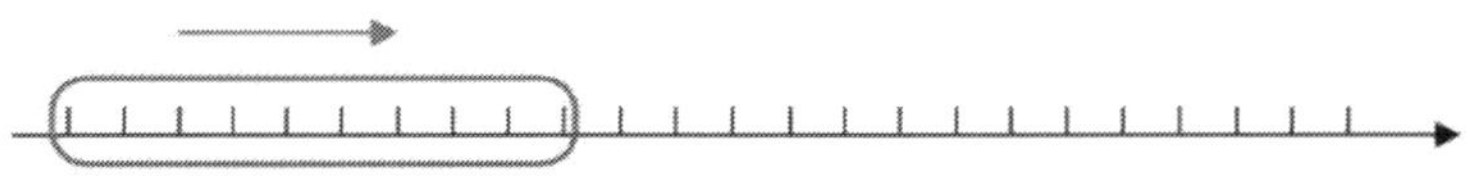

Fig.3. Window sliding schematic diagram

3.3 Hybrid energy storage capacity allocation process based on low-pass filtering and volatility observation

The configuration flow of hybrid energy storage system proposed in this paper is shown in Figure 4.

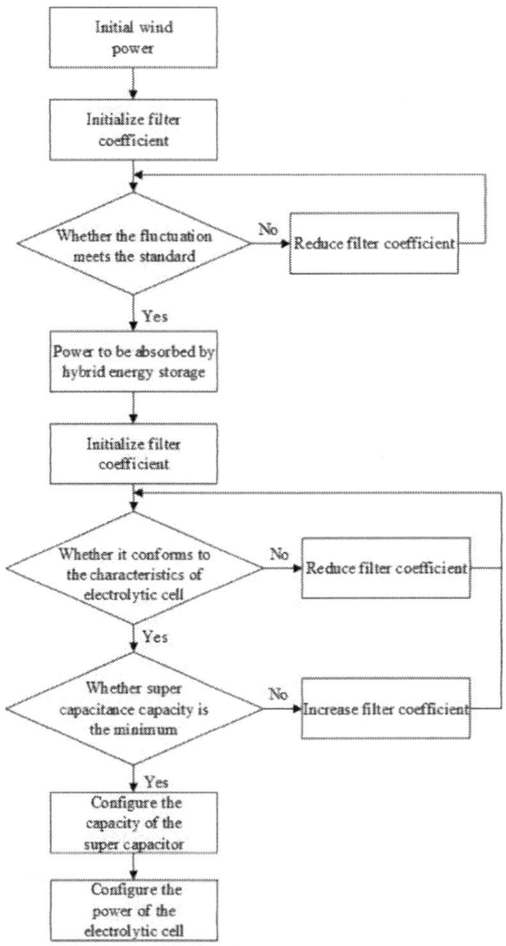

Fig.4. Hybrid energy storage configuration flow chart

The first step is to distribute wind power based on low-pass filtering. After the first low-pass filtering of the initial wind power, the new grid connected power and the fluctuating power needed to be absorbed by the energy storage system are obtained, the filtering coefficient is determined by seeking the maximum value as far as possible under the condition that the new grid connected power meets the national grid connected standard.

The second step is to distribute the absorbed power of each hybrid energy storage device. Based on the low-pass filtering method, the high-frequency fluctuating power absorbed by the super capacitor and the low-frequency fluctuating power absorbed by the electrolytic cell are obtained. The determination of the filtering coefficient needs to be determined according to the characteristics that the electrolytic cell can only absorb power and the minimum capacity of the super capacitor to achieve the minimum initial investment goal.

The third step is to configure hybrid energy storage. According to the determined filter coefficient, the theoretical power absorption curve of the super capacitor is obtained, and the capacity is configured according to the boundary conditions of the super capacitor, and the power of the electrolytic cell is configured considering the coordinated control strategy of hybrid energy storage.

4 Coordinated control strategy of hybrid energy storage system

The control strategy between cell and super capacitor mainly focuses on the power distribution between them. Starting from the state of charge of the super capacitor, the power distribution of the electrolytic cell is based on the state of charge [19]:

$$P_{el} = \frac{2P_{max}}{SOC_{max} - SOC_{min}}(SOC_t - \frac{SOC_{max} + SOC_{min}}{2}) \tag{13}$$

Where P_{max} is the overall real-time power command of the hybrid energy storage system, whose value is greater than zero indicates the energy absorbed by the energy storage system, otherwise indicates the energy released; P_{el} indicates the power of the electrolyzer; P_{sc} indicates the power of the supercapacitor; SOC_t indicates the real-time state of charge of the supercapacitor; SOC_{max} and SOC_{min} are the upper and lower limits of the state of charge of the super capacitor respectively.

It can be seen from equation (13) that when the state of charge of the super capacitor is more than half, the electrolytic cell starts to work; and when the state of charge of the super capacitor is close to the upper limit value, all the power is absorbed by the electrolytic cell; especially, because the electrolytic cell cannot output power, when the P_{max} is negative, all the power is provided by the super capacitor. The multi condition control strategy of hybrid energy storage system is as follows:

Case 1: when the energy storage system absorbs energy and the state of charge of the super capacitor is greater than 50%, that is $P_{max} > 0$， $\mathrm{SOC} > 50\%$.

 https://doi.org/10.1051/e3sconf/202015501003

$$\begin{cases} P_{el} = \dfrac{2P_{\max}}{SOC_{\max} - SOC_{\min}}(SOC_t - \dfrac{SOC_{\max} + SOC_{\min}}{2}) \\ P_{SC} = P_{\max} - P_{el} \end{cases} \tag{14}$$

Case 2: the energy storage system absorbs energy, and the state of charge of the super capacitor is less than 50%, that is $P_{max} > 0$, $\mathrm{SOC} < 50\%$.

$$\begin{cases} P_{el} = P_{T\min} \\ P_{SC} = P_{\max} - P_{T\min} \end{cases} \tag{15}$$

Case 3: the energy storage system release energy, that is $P_{max} < 0$.

$$\begin{cases} P_{el} = P_{T\min} \\ P_{SC} = P_{\max} - P_{T\min} \end{cases} \tag{16}$$

In case 2 and case 3, the electrolyzer works under the condition of thermal insulation power P_{Tmin}, while in case 2, the positive value of P_{sc} indicates that the super capacitor absorbs power from the grid, while in case 3, the negative value of P_{sc} indicates that the super capacitor releases power to the grid, so their physical meanings are different.

The transfer of three working conditions is shown in Figure 5.

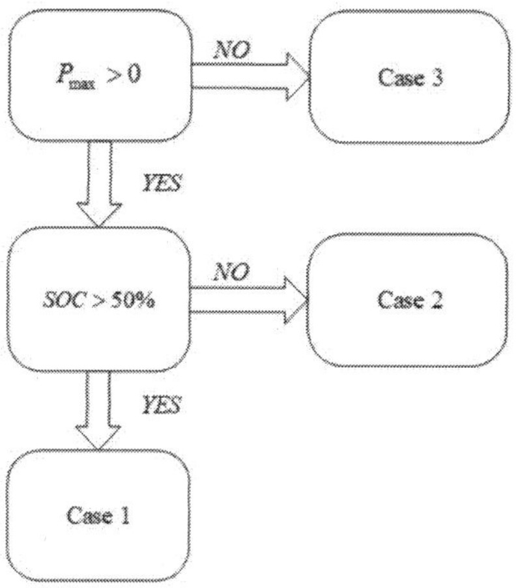

Fig.5. Working condition transfer diagram

5 Case analysis

5.1 Simulation parameter setting and wind power waveform

In order to verify the effectiveness of the proposed method, this paper takes 50MW wind farm as an example for hybrid energy storage configuration. Based on MATLAB / Simulink, build wind speed model [20] and wind turbine model [21], and the simulation parameters are shown in Table 2.

Table 2. Simulation parameters

$\boldsymbol{V_b}$(m/s)	4	ρ（kg/$\boldsymbol{m^3}$）	1.225
$\boldsymbol{V_{gmax}}$(m/s)	5	$C_p(\lambda,\beta)$	0.45
$\boldsymbol{V_{rmax}}$(m/s)	2	R(m)	43.5

Where V_b is the basic wind, V_{gmax} is the gust, V_{rmax} is the gradual wind, ρ is the air density, C_p is the wind energy utilization rate, and R is the radius of the wind turbine. The wind speed simulation model is divided into basic wind, gust, gradual wind and random wind. The accumulation of four kinds of wind can get the natural wind. After the natural wind passes through the wind turbine model, the original wind power is obtained. Through simulation, the wind speed waveform and corresponding power waveform of 10h can be obtained as shown in Figure 6.

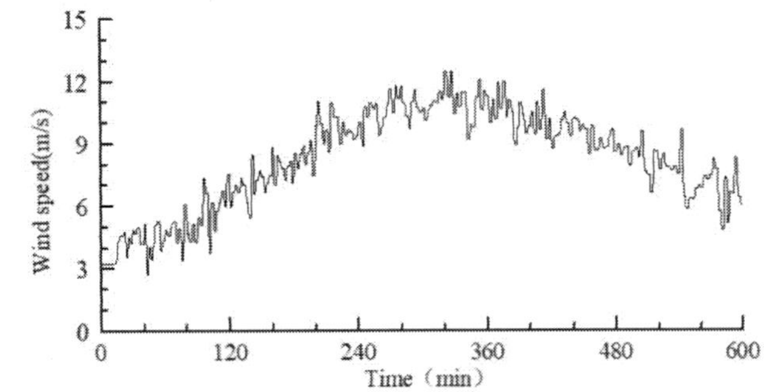

(a) Wind speed waveform

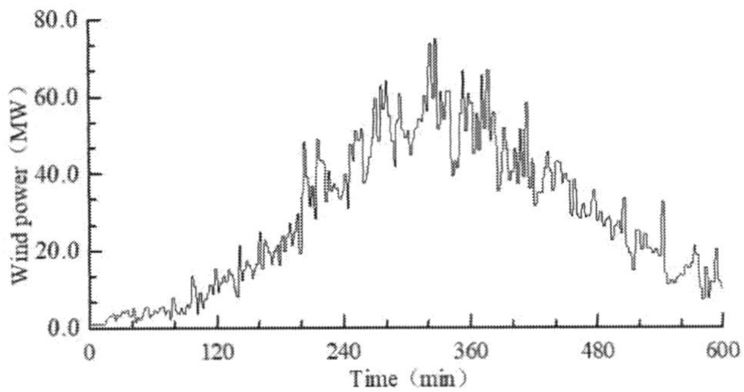

(b) Wind power waveform

Fig.6. Simulated wind speed and wind power curve

5.2 Analysis of original wind power fluctuation

Figure 6 (b) is the original wind power curve. Based on the window sliding data processing method, we can get the dispersion diagram of the amplitude and time of the fluctuation in 1min and 10min, as shown in Figure 7.

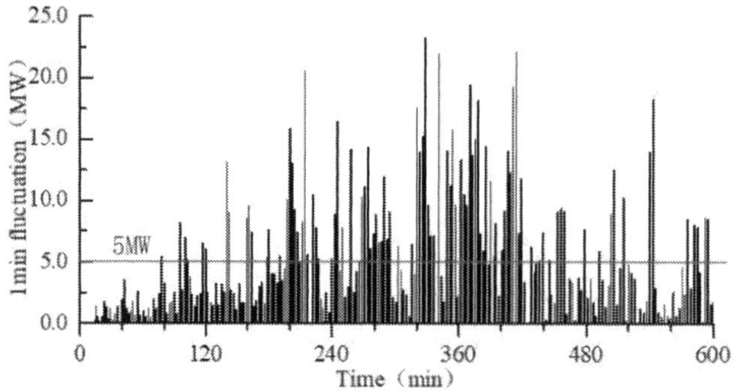

(a) 1min fluctuation without energy storage system

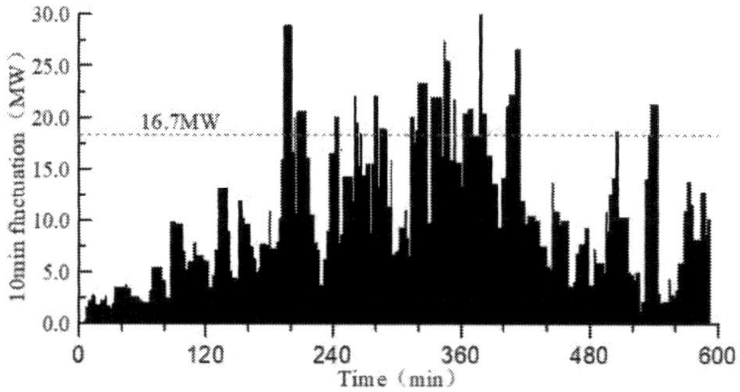

(b) 10min fluctuation without energy storage system

Fig.7. Fluctuation without the energy storage system

The red line in Figure 7 is the maximum fluctuation value allowed by the national standard, in which the maximum allowable fluctuation value of 1min is 5MW, and the maximum allowable fluctuation value of 10min is 16.7mw. Obviously, most of the fluctuation of wind power exceeds the standard before the energy storage system is added.

5.3 Hybrid energy storage power and capacity economic configuration

When the original wind power is filtered for the first time, and the filter coefficient is adjusted gradually to make the filtered grid connected power meet the national grid connected standard, then the power consumed by the energy storage system can be obtained. In order to determine the power and capacity of different energy storage equipment, the second low-pass filtering is carried out for the power that needs to be absorbed by the energy storage system, and the theoretical absorption power curve of the super capacitor is obtained. Then the capacity of the electrolytic cell is obtained by the control strategy. Under the condition of ensuring the system's stabilizing ability, the larger the capacity of the electrolytic cell is, the lower the initial investment cost of the system is. That is to say, the absorption power curve of the electrolytic cell is just positive by adjusting the second filter coefficient. At this time, the power absorption curve of the super capacitor is shown in Figure 8, and the capacity of the super capacitor configured is the minimum value W_{SCmin}, and the corresponding cost of the super capacitor is the minimum F_{SCmin}.

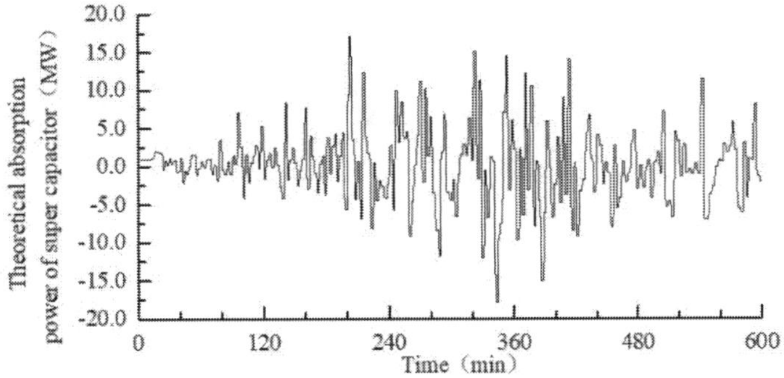

Fig.8. Theoretical absorption power waveform of super capacitor

In this paper, a discrete energy calculation method is used to determine the capacity allocation of the super capacitor. Firstly, sample and process the sample data in Fig. 8, and when $t_k = kt(k = 1,2,3\cdots n)$, calculate the real-time energy value E_{sc} in kwh according to equation (17).

$$E_{SC}(t_k) = E_{SC}(t_{k-1}) + \frac{P_{SC}(t_k)}{3.6\times10^6} \tag{17}$$

After the real-time energy data value is obtained according to equation (17), the power change curve of the super capacitor in the whole process can be drawn, from which the maximum value and the minimum value can be found, and then the super capacitor capacity to be configured can be obtained. Considering the charge state range of the super capacitor is 0.05 ~ 0.95, the final configuration capacity can be obtained by dividing the difference value by 0.9.

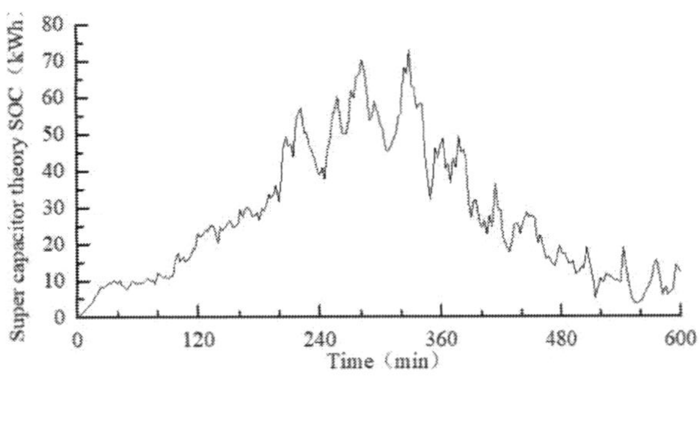

Fig.9. Super capacitor theory energy change value

According to figure 9, the difference between the maximum and minimum value is 72.95kwh. Considering the range of state of charge, the capacity is finally configured as 81.06kwh. After determining the capacity of the super capacitor, considering the coordinated control strategy between different energy storage systems, the configuration of the electrolytic cell is as follows:

According to the above control strategy, the super capacitor is charged first, and the absorption power of the electrolytic cell is determined according to the change of the state of charge, so as to obtain the power absorption curve of the electrolytic cell and the actual power absorption curve of the super capacitor in the whole process, as shown in figure 10. It should be noted that for the sake of description, the state of zero absorption power of the cell in Figure 10 is that the cell operates under the condition of thermal insulation power by default.

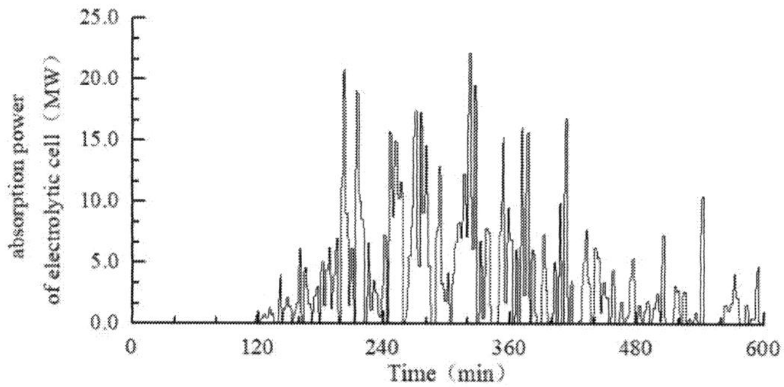

(a) Power absorption curve of electrolyzer

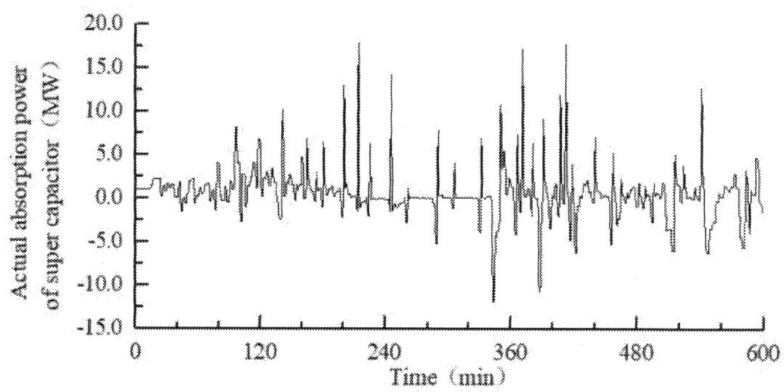

(b) Power absorption curve of super capacitor

Fig.10. Electrolyzer and Super capacitor power absorption curve

According to figure 10 (a), the capacity of the electrolytic cell is 20MW, and the corresponding cost is the maximum F_{ELmax}. As the main energy storage device, it absorbs most of the fluctuating power of wind power. As the auxiliary energy storage device, the super capacitor absorbs the high-frequency component of wind power and provides the ability to output power to the grid. Figure 11 shows the actual storage capacity of the super capacitor. It can be seen that its state of charge remains stable after saturation, ensuring the power output capacity of the energy storage system to the grid in the low wind power period.

 https://doi.org/10.1051/e3sconf/202015501003

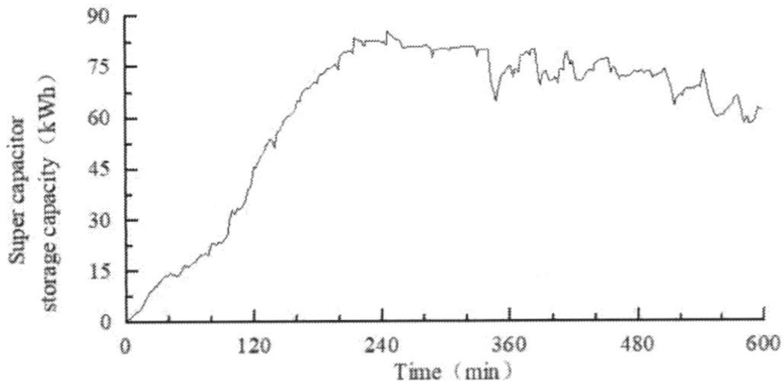

Fig.11. Super capacitor actual storage power change

5.4 Configuration result analysis

After the energy storage system is added, that is, after the low-pass filtering method is adopted, the wind power is divided into two parts: high-frequency and low-frequency, that is, the power that the energy storage system needs to absorb and the grid connected power. The grid connected power is shown in Figure 12. Compared with the original wind power, the added energy storage system can effectively reduce its volatility.

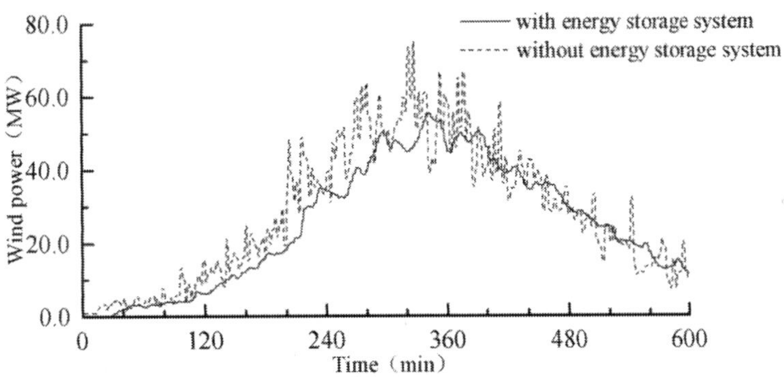

Fig.12. Grid-connected power waveform with or without energy storage system

Figure 13 is the evaluation result of the fluctuation of the stabilized grid connected power based on the window sliding observation method. According to figure 13, it is not difficult to find that the maximum value of 1min fluctuation is 3.7MW, which is less than 5MW of the national standard; the maximum value of 10min fluctuation is 9MW, which is less than 16.7MW of the national standard; at this time, the configured capacity is 20MW of electrolytic cell and 81.06kWh of super capacitor. Based on the configuration method proposed in this paper, the electrolyzer super capacitor hybrid energy storage system can effectively suppress the power fluctuation of wind power and achieve the expected goal.

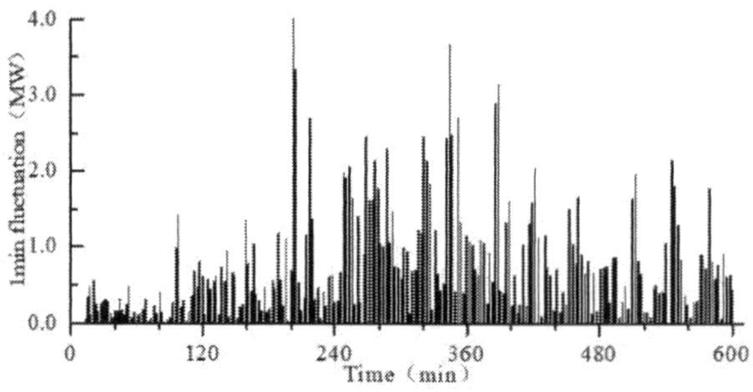

(a) 1min fluctuation of grid connected power of energy storage system

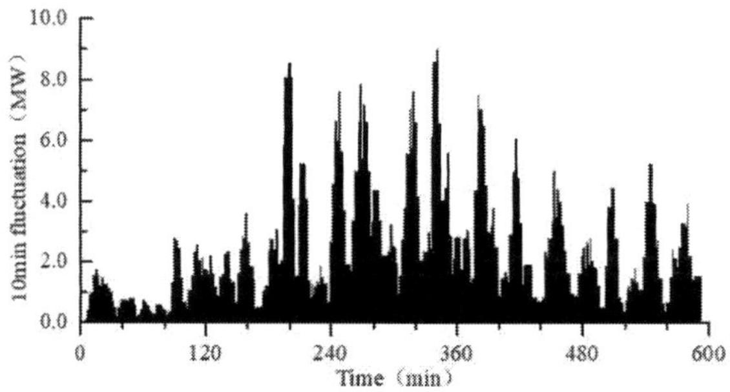

(b) 10min fluctuation of grid connected power of energy storage system

Fig.13. Fluctuation with the energy storage system

5.5 Economy analysis

First of all, for the wind farm, adding energy storage system will increase the disposable investment of the whole system, but it can make the output of the wind farm more stable and easier to connect to the grid. Practical experience shows that the energy storage system will increase the cost of wind farm by 3%, but it can reduce 20% of the abandoned wind power.

In addition, the selection of energy storage equipment in the energy storage system is also an important factor affecting the economy. Compared with the traditional battery and super capacitor hybrid energy storage system, the combination of the electrolyzer and super capacitor hybrid energy storage system requires higher capacity of the super capacitor. This is because the electrolyzer can only absorb power but not release power, so the traditional energy storage system has more advantages in terms of disposable investment; for energy conversion efficiency, battery is the mutual conversion of electric energy [22]. The energy conversion efficiency of lead-acid battery is 80%, while that of lithium-ion battery is 97%. The energy conversion efficiency of the electrolyzer is about 55%, which is also the disadvantage of the energy storage system.

However, compared with the battery, the cell has longer cycle life and lower maintenance cost, which is a better choice in the long run. More importantly, hydrogen is a kind of clean energy carrier with multi-purpose and high energy density. At present, hydrogen is mainly produced by chemical methods, which not only costs a lot but also pollutes the environment. If we use renewable energy to produce hydrogen, we can solve this problem well and make great contribution to the popularization of hydrogen energy. From this point of view, the energy storage system of electrolyzer has a more promising future.

6 Conclusion

As a solution to reduce wind power fluctuation, the wind hydrogen coupling system can not only improve the wind power and user friendliness, but also realize the green production of hydrogen, which has a good application prospect. The main conclusions are as follows:

1) A window sliding observation method of wind power fluctuation is proposed. By combining with low-pass filtering method, the quantitative evaluation of wind power fluctuation is realized intuitively.
2) Based on the characteristics of electrolyzer and super capacitor, a low-pass filter configuration method and flow chart of hybrid energy storage system are proposed. This method can realize the minimum initial investment cost under the condition of meeting the national grid connection standard.
3) The rotation optimal control strategy of the electrolytic cell and the coordinated control strategy of the cell and super capacitor hybrid energy storage system based on the SOC of super capacitor are proposed. The work of cell array equalization is realized and the power output capacity of super capacitor to power grid is guaranteed.
4) The simulation results show that the proposed hybrid energy storage system configuration method and control strategy are effective, which can effectively reduce the fluctuation of wind power grid connected power to meet the grid connected standards.

References

1. M. Ni, M. K. H. Leung, K. Sumathy, E. C, **18(5)**, 5-9, (2004)
2. M. Korpas, A. T. Holen, IEEE Trans Energy Convers, **21(3)**, 742-749, (2006)
3. Z. Z. Chen, E.E, **1**, 12-14, (2001)
4. C. X. Lü, X. R. Li, L. H. Hu, Automat Electron Power Sys, **2**, 21-29, (2015)
5. L. Lin, L. Zhu, R. Yang, Y. Gao and Q. Wu, ICPRE, 61-65, (2017)
6. G. W. Cai, C. Chen, L. G. Kong, Power Sys Techno, **40(10)**, 2982-2990, (2016)

7. X. F. Men, Y. Luo, O. D. Crisalle, Renew Energy, **34(11)**, 1595-1602, (2016)
8. T. Zhou, B. Francois, IEEE Trans Ind Electron, **58(1)**, 95-104, (2010)
9. J. C. Zhao, J. Chen, Y. X. Wang, PCRC, **38(6)**, 125-129, (2017)
10. G. W. Cai, L. G. Kong, A. Z. Xu, T China Electro Soc, **20,** (2017)
11. G. W. Cai, L. G. Kong, Y. Xue, Automat Electron Power Sys, **21**, 127-135, (2014)
12. J. Lei, Q. Gong, *International Journal of Energy Research,* (2018)
13. X. J. Shen, X. Y. Zhang, T. T. Lie, Int J Energ Res, **42(10)**, 3244-3257, (2018)
14. S. M. Schoenung, W. V. Hassenzahl, (2007)
15. Y. Z. Lei, W. S. Wang, Y. H. Yin, Pro CSEE, **22(5)**, 32-35, (2002)
16. GBT_19963-2011, *technical regulations for wind farm access to power systems*
17. D. W. Huang, G. W. Cai, D. F. Yang, et al. Acta Energi Sin
18. W. B. LI, J. P. LU, B. XU, ECEP, **3**, 439-444, (2012)
19. X. Q. Huang, R. X. Ding, W. Xiao, Elec Power Cons, **35(12)**, 67-71, (2014)
20. D. D. Li, C. Chen, Pro CSEE, **25(21)**, 41-44, (2005)
21. M. Asmine, J. Brochu, J. Fortmann, IEEE Trans Power Sys, **26(3)**, 1769-1782, (2011)
22. Z. L. Zhang, Y. Y. Cai, Y. Zhang, IEEE Trans on Power Elec, **31(1)**, 1-1, (2015)

Effect of impurity on thermally self-sustained double reactor coupling hydrogen production from glycerol reforming and methanol production from carbon dioxide and hydrogen

Sasinun Thirabunjongcharoen[1] , *and Pattaraporn* Kim-Lohsoontorn[1,*]

[1]Centre of Excellence on Catalysis and Catalytic Reaction Engineering, Department of Chemical Engineering, Faculty of Engineering, Chulalongkorn University, Bangkok, 10330, Thailand

Abstract. Thermally self-sustained double reactor (TSSDR) operating without external heat source consists of dual channels for endothermic and exothermic reactions. Hydrogen (H_2) is produced from wasted glycerol by aqueous-phase glycerol reforming (APGR) at 200-250 °C and 20-25 bar while carbon dioxide (CO_2) is a by-product. Produced H_2 and CO_2 are used as raw materials for methanol synthesis (MS) at 200-250 °C and 50-80 bar. Methanol synthesis and glycerol reforming occur at inner and outer channels of TSSDR, respectively. The TSSDR is fully packed with catalyst. Generated heat of exothermic reaction is sufficient for endothermic reaction. Main products of glycerol reforming in gas phase are H_2 and CO_2 while CO and CH_4 are by-products. All products in gas phase are totally recycled as a feed stream for exothermic channel. CO and CH_4 in feed reduce CO_2 conversion and methanol yield in MS. The effect of impurities in glycerol feed stream also influences with hydrogen production in APGR. Especially, methanol, which is an impurity in glycerol feed obtained from biodiesel production, significantly reduces glycerol conversion in TSSDR.

1 Introduction

Methanol is an essential chemical and primary feedstock for paraffins, olefins and various organic compounds such as acetic anhydride, acetic acid and formaldehyde [1]. MS is generally carried out in gas phase at high pressure and temperature of 50-80 bar and 200-300 °C with $Cu/ZnO/Al_2O_3$ as a catalyst. As this reaction is exothermic, researchers have devised schemes to use heat released from MS in other reactions. Rahmanifard et al. [2] studied MS in thermally coupled membrane where MS is a heat source for cyclohexane dehydrogenation and Nimcar et al. [3] utilized the exothermic heat from MS in benzene production. Generally, methanol is produced from reformed natural gas (syngas), resulting in CO_2 emission which has negative impact to environment. Methanol can be produced from CO_2 conversion which can be a promising method to reduce CO_2 emission while producing higher value products. However, hydrogen (H_2) source and availability are still a major barrier for CO_2 conversion.

* Corresponding author: pattaraporn.k@chula.ac.th

Glycerol is waste from biodiesel production and can be available source for H_2 production. Normally, glycerol reforming is endothermic reaction is generated in gas phase at high temperature (400-700 °C) [4]. Glycerol reforming can be carried out at different operating condition-autothermal glycerol reforming, photo-reforming and aqueous phase glycerol reforming (APGR). The features of APGR are that no vaporization of the feedstock is required, which could decrease the input energy compared to steam reforming. This reaction can operate at low temperature (200 – 250 °C) [5, 6]. Commercial catalysts for APGR is noble metal on alumina or carbon. Pt/Al_2O_3 can be used as a catalyst and was reported to provide glycerol conversion around 50-60 % [7] with a feed glycerol to water ratio of 1:9. However, production of H_2 from glycerol produces CO_2 as by-product. Direct utilization of CO_2 with H_2 obtained from APGR can be another promising pathway for glycerol conversion with mitigating CO_2 emission. APGR's products in gas phase consist of H_2, CO_2, CH_4 and CO [8]. Methanation and reverse water gas shift reaction (RWGS) is likely occurred in MS at low temperature [9].

In this study, a novel thermally self-sustained double reactor (TSSDR) coupling endothermic APGR with exothermic MS is proposed. Heat transfer between the two reactions is enabled in this reactor and lead to more favourable conditions for higher reactant conversion. H_2 and CO_2 produced from APGR can be totally utilized in MS for methanol production. A computational fluid dynamic (CFD) model was established to study this TSSDR. Effects of impurities in APGR and MS were investigated.

2 Model description

The TSSDR in our study consists of two concentric channels: 1) an inner tube and 2) an outer annular channel. Fig. 1 shows a schematic diagram of this TSSDR. The diameter of the inner channel is 16 mm and the thickness of the annulus channel is 8 mm. The wall between the two channels is 2 mm thick. The reactor height is 150 mm. The inner channel is intended for MS and the outer channel is designed for APGR. $Cu/ZnO/Al_2O_3$ catalyst for MS is packed in the inner channel, through which CO_2 and H_2 are supplied. A Pt/Al_2O_3 catalyst for APGR is packed in the outer channel. Generated heat from the exothermic reaction in the MS channel is expected to transfer to the endothermic APGR channel. Both feed streams are fed vertically from the bottom to the top of the reactor as a base case configuration.

 https://doi.org/10.1051/e3sconf/202015501004

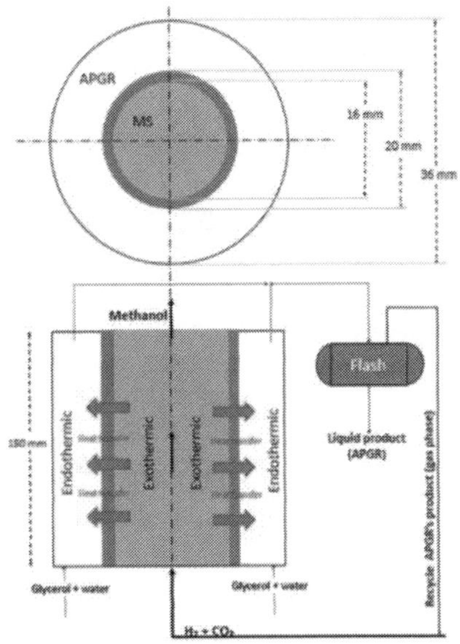

Fig. 1. TSSDR schematic design

2.1 Kinetic rate

2.1.1 Methanol synthesis (MS)

To calculate TSSDR model, methanol is synthesized by CO_2 hydrogenation. Busshe et al. [10] explains the kinetic rate of MS with $Cu/ZnO/Al_2O_3$ at 200-250 °C. In methanol production, CH_4 and CO are undesired products in process. Possible reactions in MS are shown in Table 1.

Table 1. Possible reactions and kinetic rates in MS [10,11]

Possible reaction	Rate expression
CO_2 hydrogenation: $CO_2 + 3H_2 \leftrightarrow CH_3OH + 2H_2O$	$R_{MS} = \frac{(k_1 P_{H_2} P_{H_2O}) - (k_6 P_{CO_2} P_{CO} / P_{H_2O}^2)}{(1 + (k_2 P_{CO_2} P_{H_2O}) + (k_3 P_{H_2O}^{0.5}) + (k_4 P_{CO_2}))^3}$
CO hydrogenation: $CO + 2H_2 \leftrightarrow CH_3OH$	$R_{MS,2} = k_{1,0}(P_{CO} P_{H_2}^2 - \frac{P_{CH_3OH}}{K_1})$
Methanol dehydration: $2CH_3OH \leftrightarrow CH_3OCH_3 + H_2O$	$R_{DME} = k_{2,0}(P_{CH_3OH}^2 - \frac{P_{CH_3OH} P_{H_2O}}{K_2})$
Reverse water gas shift (RWGR): $CO_2 + H_2 \leftrightarrow CO + H_2O$	$R_{RWGS} = \frac{(k_5 P_{H_2}) - (k_7 P_{CO_2} P_{CH_3OH} / P_{H_2O})}{(1 + (k_2 P_{CO_2} P_{H_2O}) + (k_3 P_{H_2O}^{0.5}) + (k_4 P_{CO_2}))}$
Methanation: $CO + 3H_2 \leftrightarrow CH_4 + H_2O$	$R_{methane} = k_{3,0}(P_{CO} P_{H_2}^3 - \frac{P_{CH_4} P_{H_2O}}{K_3})$

2.1.2 Aqueous-phase glycerol reforming (APGR)

Glycerol and water were feedstock of APGR for hydrogen production with CO_2 as a by-product. Iliuta et al. [12] formulated a kinetic model of APGR with a Pt/Al_2O_3 catalyst in a trickle bed reactor. Arely et al. [8] investigated the APGR in a batch slurry reactor. Methane, ethane and propane were detected as gaseous by-products while ethylene glycol and alcohols were detected as liquid products. The rate for side reactions in APGR are show in Table 2 when $k_{i,0}$ is calculated by Arrhenius's equation.

Table 2. Possible reactions and kinetic rates in APGR [8, 9]

Possible reaction	Rate expression
$C_3H_8O_3 + 3H_2O \rightarrow 3CO_2 + 7H_2$	$R_{APGR,1} = 2.678\exp\left(-\frac{60000}{RT}\right)C_{C_3H_8O_3}$
$C_3H_8O_3 \rightarrow 3CO + 4H_2$	$R_{APGR,2} = k_{2,0}\exp\left(\frac{-83990}{RT}\right)C_{C_3H_8O_3}$
$CH_3OH + H_2 \rightarrow CH_4 + H_2O$	$R_{APGR,3} = k_{3,0}\exp\left(\frac{-84510}{RT}\right)C_{CH_3OH}\frac{C_{H_2,g}}{H_{H_2}}$
$CO + H_2O \rightarrow CO_2 + H_2$	$R_{APGR,4} = k_{4,0}\exp\left(\frac{-130300}{RT}\right)\frac{C_{CO_g}}{H_{CO}}$
$CO + 3H_2 \rightarrow CH_4 + H_2O$	$R_{APGR,5} = k_{5,0}\exp\left(\frac{-133260}{RT}\right)\frac{C^{1.5}_{H_2,g}}{H_{H_2}}$
$CH_3OH \rightarrow CO + 3H_2$	$R_{APGR,6} = \frac{(1.17 \times 10^{-7}\exp(\frac{-40999}{RT}))C_{CH_3OH}}{1 + (5.48 \times 10^{-5})C_{CH_3OH}}$
$3CO_2 + 4H_2 \rightarrow CH_4 + 2H_2O$	$R_{APGR,7} = (1.75 \times 10^{-7})\exp\left(\frac{-95000}{RT}\right)C^{0.66}_{CO_2}C^{0.21}_{H_2}$

2.3 Simulation method

MS and APGR in TSSDR were simulated with COMSOL Multiphysics 5.3a. Steady-state flows of the feed gas mixtures was solved by mass, momentum, and energy balance equations of a discretized flow domain of the reactor geometry. The dual reactions were simulated in a 2D-axis-symmetric system in cylindrical coordinate. The flow regime in each channel was assumed to be steady and laminar. The diffusion of materials and the heat conduction in direction flow axis was neglected. The ideal gas and Peng-Robinson model was employed. The physical properties of the chemicals were initially defined with Perry's Chemical Engineers' Handbook [13] . The mathematical equations for the CFD model were divided into three parts. For Part I, MS was calculated at the surface of the solid catalyst. A constant bed void fraction was maintained for the entire length of the reactor. For Part II, APGR was calculated with three phases (gas-phase products, liquid-phase reactants and solid catalyst). For Part III, heat transfer through the solid wall was determined. Numerical solutions were obtained with the finite element method in COMSOL Multiphysics version 5.3a. The mesh resolution was free triangular with a mesh area of 0.0012 m^2 and the number of elements of 1416. Performance of both reactions are measured by conversion, yield and selectivity, show in Eq. (1-6).

$$CO_2\ conversion = \frac{mol\ of\ CO_2\ inlet - mol\ of\ CO_2\ outlet}{mol\ of\ CO_2\ inlet} \times 100 \quad (1)$$

$$C_3H_8O_3\ conversion = \frac{mol\ of\ C_3H_8O_3\ inlet - mol\ of\ C_3H_8O_3\ outlet}{mol\ of\ C_3H_8O_3\ inlet} \times 100 \quad (2)$$

 https://doi.org/10.1051/e3sconf/202015501004

$$\%\,yield\ of\ H_2 = \frac{mol\ of\ H_2\ generated}{mol\ of\ C_3H_8O_3\ feed} \times 100 \tag{3}$$

$$\%\,yield\ of\ CH_3OH = \frac{mol\ of\ CH_3OH\ generated}{mol\ of\ CO_2\ feed} \times 100 \tag{4}$$

$$\%\,selectivity\ of\ APGR\ product = \frac{mol\ of\ desired\ product}{mol\ of\ C_3H_8O_3\ reacted} \times 100 \tag{5}$$

$$\%\,selectivity\ of\ MS\ product = \frac{mol\ of\ desired\ product}{mol\ of\ CO_2\ reacted} \times 100 \tag{6}$$

3 Result & Discussion

Simulation for the model validation was performed for the MS and APGR channels separately in adiabatic model. The MS model accuracy was assured at an average error of only 0.88 % based on average of all components when compared with Busshe et al [10]. In APGR, the average error is 2.82% when comparing output composition with Iliuta et al. [12]. This research divides TSSDR model into three cases. Case I is the base case condition: MS inlet temperature of 220 °C; MS inlet pressure of 50 bar; MS total flow rate of 0.003 mol s^{-1}; MS external feed H_2 to CO_2 ratio of 82:3; APGR inlet temperature of 220 °C ; APGR inlet pressure of 20 bar; MS total flow rate of 0.0005 mol s^{-1}; APGR external feed glycerol to water ratio of 1:9. Case II is the TSSDR model without external feed of H_2 and CO_2 in MS feed stream. MS feed totally relied on APGR's gas products. Case III is the TSSDR model was similar to case II plus methanol as impurity in APGR feed side.

As presented in Fig 2, methanol yield and CO_2 conversion in base case were higher than other conditions since H_2 was highly excess. CO_2 hydrogenation is based on theoretical ratio of H_2 to CO_2 at 3. There was a reaction between hydrogen and CO (CO hydrogenation). Therefore, methanol yields in base case was higher than CO_2 conversion. On the other hand, total recycle of APGR's gas product to MS influenced CO_2 conversion decreased to 36.36% because the ration of H_2 to CO_2 in feed was decreased to 3:1. Gas products in APGR consisted of the main product (H_2, CO_2) and undesired products (CO, CH_4 and methanol) which led to a reduction in methanol production.

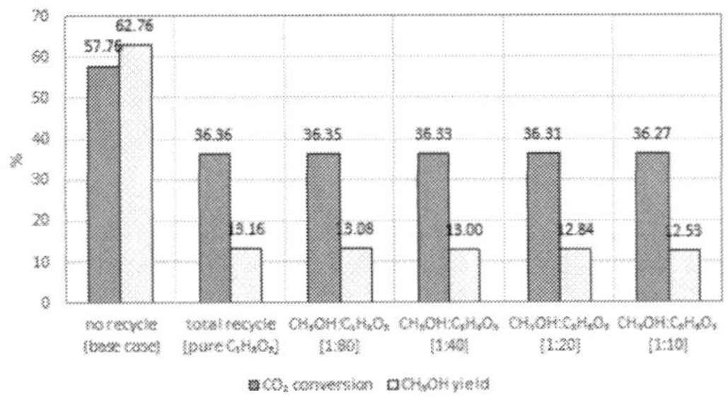

Fig. 2. Comparing CO_2 conversion and CH_3OH yield in exothermic side: (1) TSSDR with external feed H_2 and CO_2, (2) TSSDR with total recycle APGR's gas product, (3) Effect of methanol impurity in APGR's feed stream (ratio = 1:80), (4) Effect of methanol composition in APGR's feed stream (ratio = 1:40), (5) Effect of methanol composition in APGR's feed stream (ratio = 1:20), (6) Effect of methanol composition in APGR's feed stream (ratio = 1:10).

As presented in Fig. 3, there were a minor change in H_2 yield and glycerol conversion in all cases because the molar of pure glycerol in water solution was constant although methanol was introduced as impurity in feed.

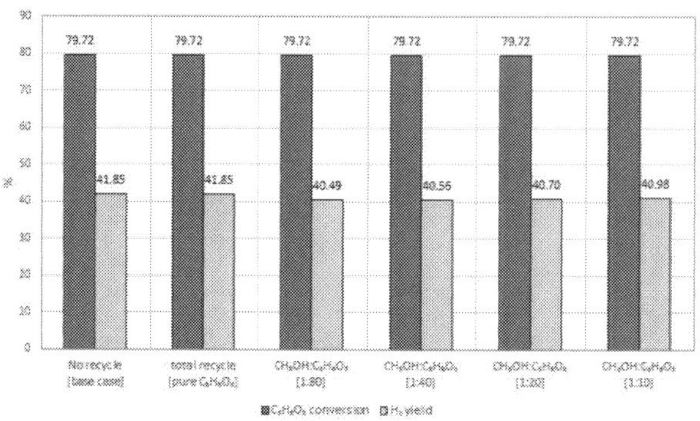

Fig. 3. Comparing $C_3H_8O_3$ conversion and H_2 yield in endothermic side: (1) TSSDR with external feed H_2 and CO_2, (2) TSSDR with total recycle APGR's gas product, (3) Effect of methanol composition in APGR's feed stream (ratio = 1:80), (4) Effect of methanol composition in APGR's feed stream (ratio = 1:40), (5) Effect of methanol composition in APGR's feed stream (ratio = 1:20), (6) Effect of methanol composition in APGR's feed stream (ratio = 1:10).

In reality, crude glycerol which is obtained from biodiesel production could contain methanol as impurity. Fig. 4 shows the effect of methanol in glycerol feed in APGR on product selectivity in MS. Selectivity of MS's product insignificantly decreased. There was a separation of liquid product by condensation before recycling to MS. Methanol was removed before entering MS.

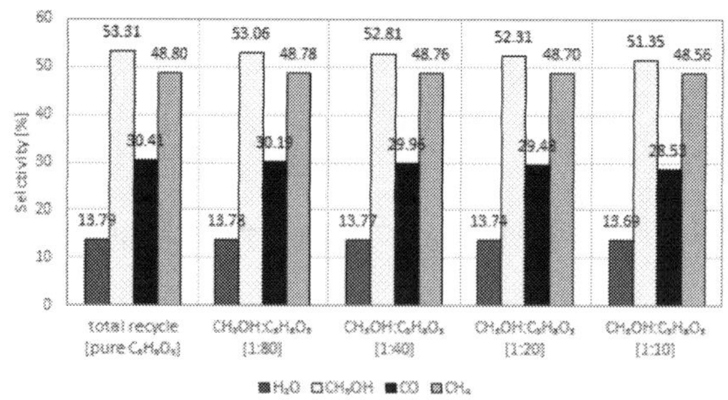

Fig. 4. Selectivity of MS's product base on CO_2 reacted: (1) TSSDR with external feed H_2 and CO_2, (2) TSSDR with total recycle APGR's gas product, (3) Effect of methanol composition in APGR's feed stream (ratio = 1:80), (4) Effect of methanol composition in APGR's feed stream (ratio = 1:40), (5) Effect of methanol composition in APGR's feed stream (ratio = 1:20), (6) Effect of methanol composition in APGR's feed stream (ratio = 1:10).

Fig. 5 presents effect of methanol impurity in APGR feed on the selectivity of APGR products. Methanol selectivity dropped to the negative value. Generated methanol is less than molar feed of methanol. Methanol could decompose to H_2 and CO and generated CH_4 following the reaction presented in Table 2. Methanol in glycerol solution decreased when the ratio of impurity increased.

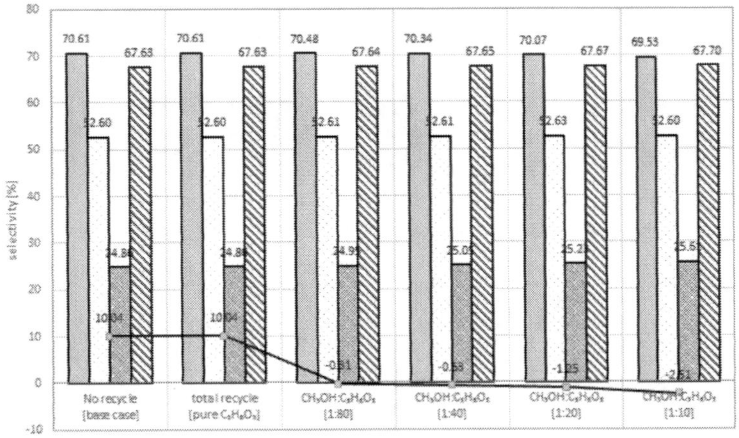

Fig. 5. Selectivity of APGR's product base on theoretical $C_3H_8O_3$ reacted: (1) TSSDR with external feed H_2 and CO_2, (2) TSSDR with total recycle APGR's gas product, (3) Effect of methanol composition in APGR's feed stream (ratio = 1:80), (4) Effect of methanol composition in APGR's feed stream (ratio = 1:40), (5) Effect of methanol composition in APGR's feed stream (ratio = 1:20), (6) Effect of methanol composition in APGR's feed stream (ratio = 1:10).

4 Conclusion

Thermally self-sustained double reactor (TSSDR) consisted of dual channels for APGR and MS. TSSDR could operate without external heat and external feed supplied in MS. H_2 and CO_2 was produced APGR and was recycled to produce methanol in MS. Therefore, by-products in APGR affected methanol production in MS. The effect of methanol in glycerol feed stream influenced hydrogen production in APGR. Therefore, feed compositions in MS and APGR were important variables and significantly affected the reactor's performance.

Acknowledgment is made to Thailand Graduate Institute of Science and Technology (TGIST) fund from National Science and Technology Development Agency (NSTDA) and CAT-REACT Industrial Project from Thailand Science and Research Innovation (TSRI). P. Kim-Lohsoontorn would like to acknowledge Research Cess Fund, Malaysia-Thailand Joint Authority (MTJA), Thailand Science Research and Innovation (TSRI) and NSTDA Chair Professor Grant from NSTDA.

References

1. A.C.C. de Souza, J.L. Silveira, Renewable and Sustainable Energy Reviews, **15**, 1835-50, (2011).
2. H. Rahmanifard, R. Vakili, T. Plaksina, Mohammad Reza Rahimpour, X.F. Masoud Babaei, Computers & Chemical Engineering, **119**, 258-69, (2018).

3. S.C. Nimkar, R.K. Mewada, M.A. Rosen, International Journal of Hydrogen Energy, **42**, 28113-27, (2017).
4. C.A. Schwengber, H.J. Alves, R.A. Schaffner, F.A.d. Silva, R. Sequinel, V.R. Bach, R.J. Ferracin, **58,** 259-6, (2016).
5. D.A. Boga, F. Liu, P.C.A. Bruijnincx, B.M. Weckhuysen, Catalysis Science & Technology, **6**, 134-43, (2016).
6. P.V. Tuza, R.L. Manfro, N.F.P. Ribeiro, M.M.V.M. Souza, Renewable Energy, **50**, 408-14, (2013).
7. V. Domínguez-Barroso, Fuel processing technology, **v. 190**, pp. 21-8-2019 v.190, (2019).
8. A. Torres, D. Roy, B. Subramaniam, R.V. Chaudhari, Ind Eng Chem Res, **49**, 10826-35, (2010).
9. I. Coronado, A. Arandia, M. Reinikainen, R. Karinen, R.L. Puurunen, J. Lehtonen, Catalysts, **9**, (2019).
10. V.B. KM, F. Gilbert, Journal of Catalysis, **161**, 1-10, (1996).
11. A.T. Aguayo, J. Ereña, D. Mier, J.M. Arandes, M. Olazar, J. Bilbao, Industrial & Engineering Chemistry Research, **46**, 5522-30, (2007).
12. I. Iliuta, M.C. Iliuta, P. Fongarland, F. Larachi, Chemical Engineering Journal, **187**, 311-27, (2012).
13. R.H. Perry, D.W. Green, *Perry's chemical engineers' handbook*, (2008).

Formation and evolution of porosity during high temperature creep of a nickel-based single crystal superalloy

Weiwei Liu[1,*], *Yuanyuan* Guo[1], *Mai* Zhang[1], and *Jian* Zhang[1]

[1]Science and Technology on Advanced High Temperature Structural Materials Laboratory, Beijing Institute of Aeronautical Materials, Beijing 100095, P.R. China

Abstract: A Re-containing single-crystal superalloy was used to research the high temperature low stress creep behavior. Transmission electron microscope, scanning electron microscope and some other research methods are employed. The results and analysis are summarized below: Two mechanisms for the steady creep are found in this experiment. The volume fraction of pores after creep test at 1100°C increased more than 2 times compared with that before test, but the increasing at 1000°C is relatively small, which reveals that temperature has an great influence on the formation of pore during creep; There are two types of pores associated with fracture during the creep process. One is the casting shrinkage located between the interdentritic, which is formed in the solidification of the alloy. Another type of pore is nucleated and growing during the creep deformation.

Key words: single crystal superalloy ; creep ; dislocation ; porosity

Nickel-based superalloys have high temperature fatigue and creep strength, and also have good thermal corrosion resistance, so they are widely used in making turbine blades

* Corresponding author: liuww_80@sina.com

and guide blades of gas turbines. Under actual service conditions, blades are subjected to high temperature and high stress. Especially during engine operation, high temperature combined with centrifugal stress will lead to creep deformation of blades, which will seriously affect the service life of blades[1,2]. With the increase of turbine temperature in advanced aeroengines, high temperature creep fracture has been one of common failure modes of single-crystal(SC) superalloy blades. Creep behaviour of Ni-based SC superalloys under typical condition has been researched extensively. In order to design and predict the service life of components reasonably, the creep behavior of superalloy needs to be fully understood and mastered. In the past decades, the creep behavior of monocrystalline superalloys has received extensive attention. The results show that the creep behavior and fracture mechanism of SC superalloy is affected by experimental condition, including temperature, loading mode, stress and other factors. With the change of these factors, the creep behaviour and deformation mechanism show different patterns [3-16].

In this paper, the high temperature creep of Ni-base SC superalloy under low applied stress at 1100℃ and 1000℃ was studied. The creep behaviour, fracture characteristics and creep deformation, especially the formation and growth of porosity in monocrystalline alloys were systematically researched and analysed. The results can provide experimental data for the application of SC alloys at high temperature and optimize the creep theory of Ni-base SC superalloys.

1. Experimental procedure

The material used in this article is a first-generation Ni-base SC superalloy containing the chemical components of Ni-Cr-W-Ta-Al-Co-Ti-Mo system. A series of constant load tensile creep experiment were carried out on a creep machine with 120 ~ 220 MPa stress. The test temperatures were 1000℃ and 1100℃ respectively. The creep strain was measured by an extensometer attached to the shoulder of the creep sample, with a measurement accuracy of 5×10^{-4}~5×10^{-5}. Most of the specimens were tensile fracture, and some of the specimens were subjected to the creep interruption experiment to observe the microstructure.

2. Results and discussion

2.1 Creep behaviour

https://doi.org/10.1051/e3sconf/202015501005

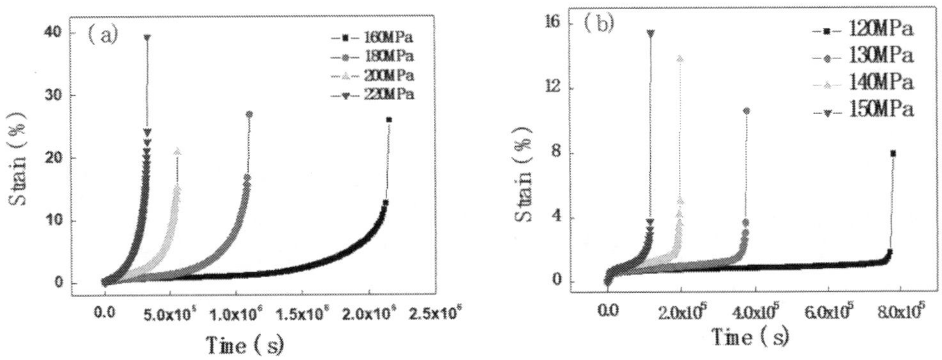

Fig. 1 Creep curves under different condition: (a) 1000 °C and (b) 1100 °C

Creep curves of the alloys at different stress at two temperatures are shown in figure 1. As can be seen from the figure, its high-temperature creep behavior mainly shows the following characteristics:

(1) Under the test conditions of 1000℃ and 1100℃, the creep curve shows the creep characteristics of three stages: initial creep(first stage), steady creep(secondary stage) and accelerated creep(third stage).

(2) When temperature increases from 1000℃ to 1100℃, the creep elongation decreases significantly. At 1000℃, the elongation is 20% ~ 40%.The elongation at 1100℃ is 8% ~ 16%, which is smaller than that of 1000℃.

(3) At 1000℃, the creep strain rate increased slowly after the alloy entered the third phase of creep. At 1100℃, the strain rate of creep increased quickly after the start of the accelerated stage, then, rupture occurred in a very short time.

2.2 Microstructural evolution

It's can be seen from Fig. 2, during the creep experiment at high-temperature ranging from both 1100 °C to 1000 °C , raft of directional coarsening of γ′ phase occurred. The microstructure of creep interruption specimen indicates that directional coarsening of γ′ phase occurs rapidly in the initial stage of creep, and raft structure has been formed in early stage of creep steady state. During the post-stage of creep test, the thickness of γ′ raft structure increases by degrees and the raft structure begin to interconnect with each other, which causes the topological inversion of γ′ phase surrounding γ matrix in the middle and post stages of creep.

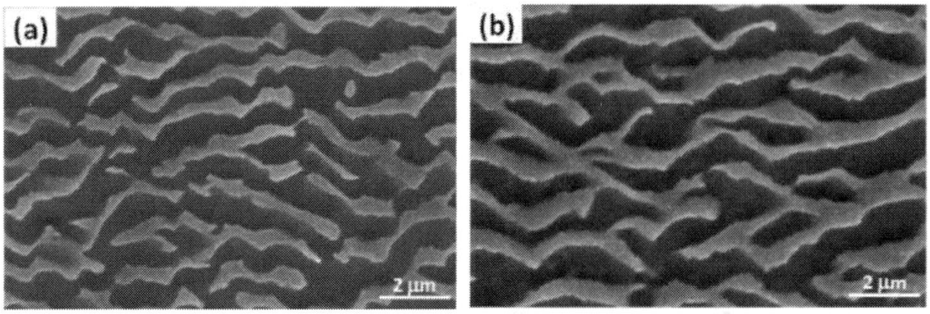

Fig. 2 Directional coarsening of γ′ phase after high-temperature creep rupture

(a) 1000℃-200MPa; (b) 1100℃-130MPa

In the experiments of 1100 °C and 1000 °C. The dislocation TEM observation shows that regular dislocation network at the γ/γ′ interface is quickly formed because of the reaction between dislocations. It is can be seen that γ′ phase was cut by superlattice dislocation in secondary stage of creep at both 1000 °C and 1100 °C (Fig. 3a and b). Which is the main deformation mechanism of steady-state creep. By comparison, there are much more dislocations cutting into γ′ phase. However, there are only a few dislocations cutting into γ′ phase under 1100 °C and lower stress, and show a single form of screw dislocations. Which is main difference between 1000°C and 1100 °C test on the creep deformation mechanism.

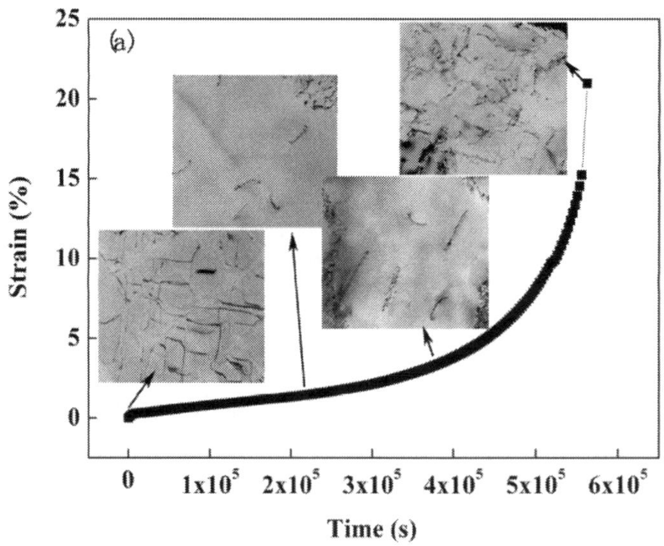

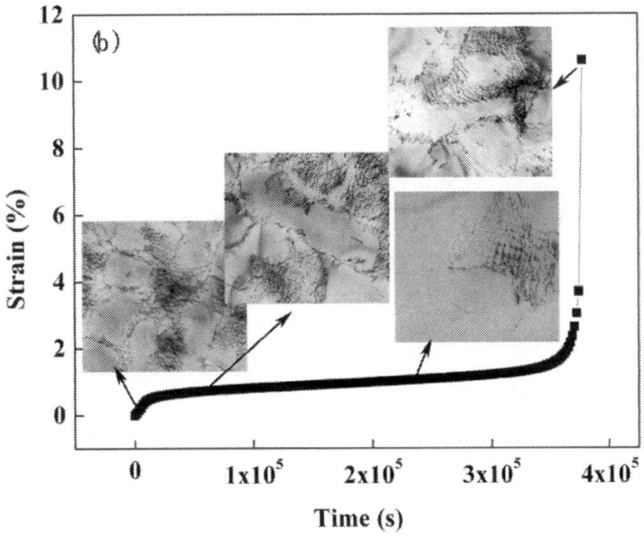

Fig. 3 Combination of creep curves with microstructural evolution during the creep

(a) 1000℃-200MPa ; (b) 1100℃-130MPa

2.3 Behaviour of porosity and creep rupture

As seen from Fig. 4 , the fracture morphology features after creep rupture in this test show slight different between 1000 °C and 1100 °C. There are abundant square facets on creep fracture surface, the orientation of planar edge is along <110> direction. Small circular holes can be seen at the center of planes, and which are the origin of cracks. In tests of 1100°C, the border line of square facets in the fracture gradually becomes irregular and unsharp, the number of crack initiation points strikingly increases.

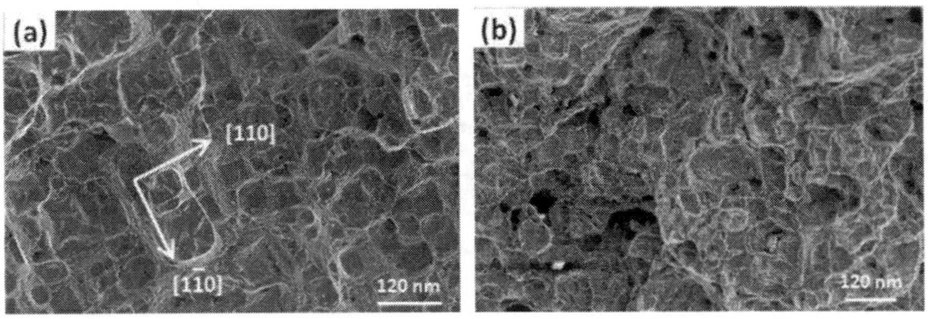

Fig. 4 Fracture morphology of after creep rupture

Longitudinal microstructure morphology near the fracture after creep test presented in Fig. 5 shows that a large number of micro-cracks exist in the necking zones near the fracture. The initiation and growth of these micro-cracks are the main controlling factors of alloy fracture. The main defects in SC alloy are casting shrinkage in the interdendritic region. It can also be seen from Fig. 5 that the main origin of micro-cracks is casting pores in the interdendritic region under the condition of 1000°C and 200 MPa. However, the number of micro-cracks is obviously more than that of the former under the condition of 1100°C and 130 MPa. Further observation shows that micro-cracks are distributed throughout the interdendritic and dendritic regions. This indicates that there are new creep pores formed in addition to casting pores in the process of creep at 1100°C, which might induce the initiation of cracks. The comparison of the volume fraction of pores away from the fracture with that of pre-deformation alloys presented in Table 1 shows that the volume fraction of pores after creep test at 1100 °C increased more than 2 times compared with that before test, but the increasing at 1000 °C is relatively small, which reveals that temperature has an great influence on the formation of pore during creep, and verified by the observation of the longitudinal microstructure morphology under other stress conditions at 1000 °C and 1100 °C.

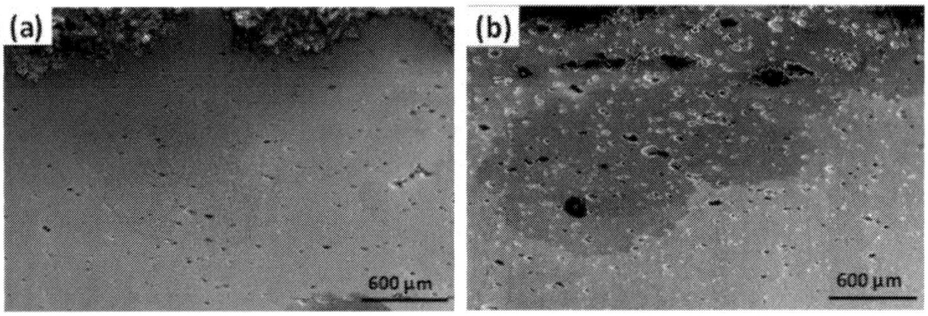

Fig. 5 Longitudinal microstructure morphology of fracture after creep fracture

Table 1 Comparison of pore fraction before and after deformation

	Undeformed	1000 °C、200 MPa	1100 °C、130 MPa
Porosity (%)	0.17	0.21	0.37

In order to investigate the high temperature creep fracture mechanism of the alloy, it is necessary to further analyze the pores that cause the crack initiation. As mentioned above, there are two types of pores associated with fracture during the creep process. One is the casting shrinkage located between the interdentritic, which is formed in the solidification of the alloy. In this work, it is defined as I-type pore. Another type of pore is nucleated and growing during the creep deformation, which is defined as II-type pore.

the variation of the I-type pore during creep is presented in Fig. 6. It can be seen that the I-type pore grows with the increase of creep deformation and gradually evolves into a polyhedral shape, and its surface orientation conforms to a certain crystallographic law. At 1100 °C, the cross-section of pore is mostly pentagonal orhexagonal, with plane paralleling to {100} plane or {110} plane, However, the cross-section of pore is mostly quadrilateral at 1000 °C, with plane paralleling to {100} plane. Combined with the analysis of specimen under other conditions, it shows that the higher the experimental temperature, the higher the degree of polyhedralization of pore and the clearer the polyhedral contour. The reason for the above phenomenon is that the surface energy has obvious anisotropy at higher experimental temperatures, which is the characteristic of pores when grown up in the form of vacancy aggregation. From the micro-fracture mechanism, the structure at the apex angle of the polyhedral pore is equivalent to a lot of micro-notches, and the stress concentration effect around the notch makes it easy to crack during creep process, especially when the specimen is necked and the stress is increased. The initiation of microcracks at the corners of the pore,and its connection with surrounding cracks after expansioncan be seen in Fig. 7. The degree of polyhedralization of pores is low at 1000°C, and the phenomenon of crack initiation at sharp corners is not obvious compared with the experiment at 1100 °C.

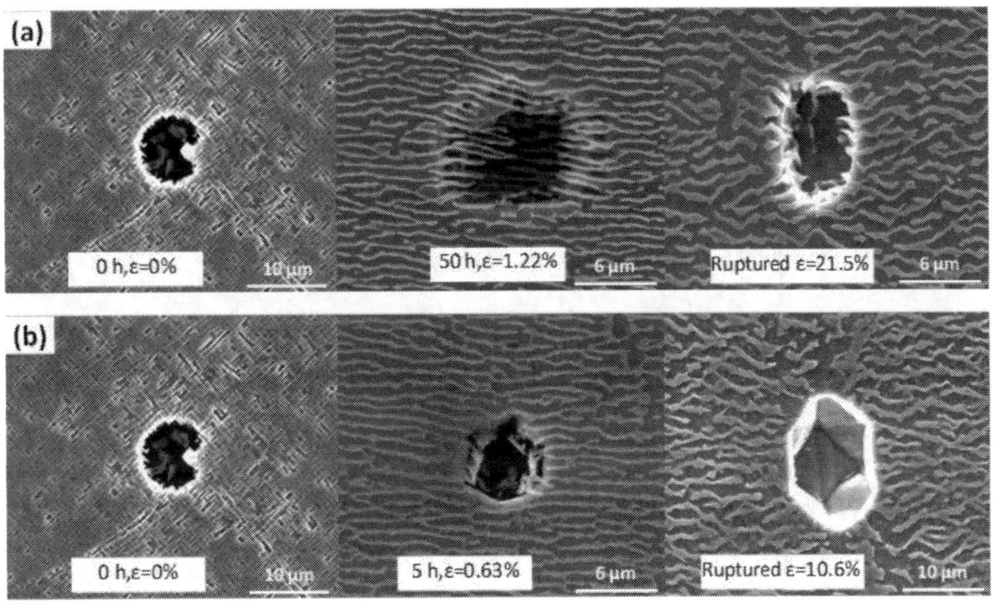

Fig. 6 Evolution of I-type pores during creep

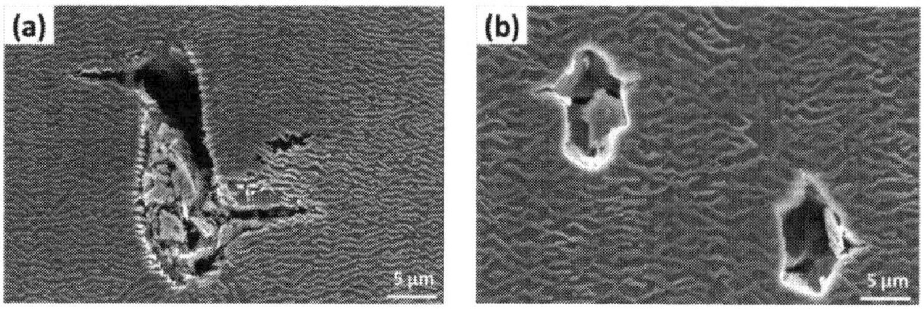

Fig. 7 cracking of I-type pore in the third stage of creep

Fig.8 shows the initiation and cracking of II-type pore during creep. II-type pores are more likely to occur at the γ/γ′ phase interface of the dendritic region and gradually grow up during the creep process. At 1000 °C, the number of II-type pores is small due to poor atom diffusion capacity, and no cracking is observed in the third stage of creep, so the impact on the creep process is minimal. In contrast, the growth trend of II-type pores at 1100 °C is obvious. In the post stage of creep, the pores in the necking region can be extended along any path in the γ/γ′ structure under stress, and together with I-type pore to cause the rapid cracking, which is also directly causing the small proportion , faster creep rateand small total elongation in thethird stage of creep test at 1100 °C.

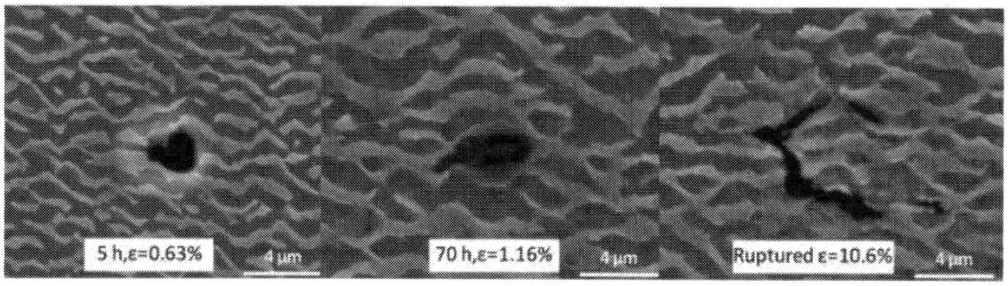

Fig.8 Formation and cracking of II-type pores during creep at 1100 °C

2.4 Analysis and discussion

The creep behavior of SC superalloys usually shows distinct deformation and fracture mechanisms according to changes in stress and temperature. In summary, under the condition of low temperature and high stress, dislocations will be cut into the γ′ phase by stacking faults, dislocation pairs or combination of both under high applied stress. With the increase of temperature and the decrease of stress level, the diffusion ability of elements in alloys gets enhanced, and the dislocations are more likely to be deformed by crossing the precipitated phase in such a way as climbing, Orowan bypassing, etc. When the experimental temperature is further increased above 1000 °C, the creep mechanism will change again. The micro-deformation characteristics of creep process in this work have a certain degree of representativeness, and the main manifestations are as follows: the movement of dislocations in the matrix and the formation of dislocation networks, superlattice dislocation cutting γ' precipitates, directional coarsening (shaped raft) of γ' phase and initiation and growth of pores.

The results of creep tests in this work show that there are more dislocations cutting γ' phase during 1000 °C creep, which is the result of the larger applied stress ranges. In the post stage of creep steady state, the coarsening of γ' phase lead s the γ/γ' dual-phase structure to be topologically inverted, the raft structures are interconnected and structurally transformed into the matrix phase, and the superlattice dislocations entering the interior of raft structures can be move continuously without interface hindrance, which causes the deformation of alloy to be intensified and accelerates the creep into the third stage. At the same time, It can be seen from the observation of volume fraction of pores and the longitudinal section of fracture, the increase of total pores at 1000 °C creep is limited, indicating that the interface dislocations climbing mechanism is suppressed at 1000 °C, and the main deformation mechanism after creep steady state is dislocations cutting γ' phase. In contrast, there is only a small amount of superlattice dislocations formed at1100 °C creep, indicating that the stress ranges of 120-150 MPa are not enough to press a large amount of dislocations into γ' phase. On the other hand, pre-existing pores in the alloy undergo significant polyhedral growth, and a large number of new creep pores appear, which means that the diffusion capacity of elements is significantly improved after the temperature raising to 1100 °C, and the interface dislocations climbing mechanism begins to become the main deformation mechanism in the middle and post stage of creep. Only few amounts of deformations occur during the whole creep steady state, which indicates that the contribution of the interface dislocations climbing mechanism to creep deformation is small, and its main influence is extruding vacancies and continuous accumulation at pores. When the cross-sectional area of pores reaches critical value, the alloy rapidly enters the third stage and fractures. The simultaneous fracture of both types of pores also causes the elongation of the alloy to be significantly reduced.

3 conclusions

In this article, a Ni-based SC superalloy was used to research the high temperature creep under low applied stress at 1100 °C, 120-150 MPa, and 1000°C, 160-220 MPa. The main findings of the research are summarized as below:

1. At 1000 °C, the accelerated stage of creep lasts longer and deformation rate changes slowly, the elongation is obviously larger than 1100°C. The deformation rate of the accelerated stage quickly increases and ruptures within a short time at 1100°C. By contrast, the total elongation decreases by around 50% compared with that of 1000 °C;

2. The volume fraction of pores after creep test at 1100 °C increased more than 2 times compared with that before test, but the increasing at 1000 °C is relatively small, which reveals that temperature has an great influence on the formation of pore during creep;

3. There are two types of pores associated with fracture during the creep process. One is the casting shrinkage located between the interdentritic, which is formed in the solidification of the alloy. Another type of pore is nucleated and growing during the creep deformation.

References

[1] R. C. REED. The superalloys: Fundamentals and applications. (Cambridge University Press, 2006)

[2] Y. TANG, M. HUANG, J. XIONG, et al. Acta Materialia, **126** (2017)

[3] S. MA, X. LV, J. ZHANG, et al. Journal of Alloys and Compounds, **743** (2018).

[4] S. ZHANG, S. G. TIAN, et al. Rare Matel Materials and Engineering, **39** (2010).

[5] Z. X. WEN, D. Zhang, S. Li, et al. Journal of Alloys and Compounds, **692** (2017).

[6] B. SHOLLOCK, J. BUFFIERE, R. CURTIS, et al. ScriptaMaterialia, **36** (2017).

[7] C. M. F. RAE, R. C. REED. Actaclia, **55** (2007).

[8] S. TIAN, X. DING, Z. GUO, et al. Materials Science and Engineering: A, **594** (2014).

[9] P. WOLLGRAMM, D. BÜRGER, A. B. PARSA, et al. High Temperature Technology, **33** (2016).

[10] M. HUANG, L. ZHUO, J. XIONG, et al. Philosophical Magazine Letters, **95** (2015).

[11] T. M. POLLOCK, A. S. ARGON. Acta Metallurgica Et Materialia, **40** (1992).

[12] H. S. M. HAFEZ, G. EGGELER, D. RAABE. Acta Materialia, **61** (2013).

[13] M. PROBST-HEIN, A. DLOUHY, G. EGGELER. Acta Materialia, **47** (1999).

[14] J. X. ZHANG, J. C. WANG, H. HARADA, et al. Acta Materialia, **53** (2005).

[15] M. HUANG, Z. CHENG, J. XIONG, et al. Acta Materialia, **76** (2014).

E3S Web of Conferences **155**, 01006 (2020) https://doi.org/10.1051/e3sconf/202015501006
HEET 2019

Feature extraction of On-Load Tap-Changer Based on Improved EMD Energy Spectrum

Huang Wenquan*

School of Information Science and Engineering, Huaqiao University, Xiamen, 361021, China

Abstract: On-Load Tap-Changer is an important device with high failure rate in power systems. This paper proposes an improved EMD energy spectrum feature extraction method based on vibration signals. It divides the entire OLTC shifting process into two sections, and extracts their characteristic energy spectrum. Analysis of sample data shows that this method can more effectively distinguish the state of OLTC.

1. Introduction

The On-Load Tap-Changer (OLTC) plays an important role in the process of the transformer adjusting the reactive power flow and stabilizing the center voltage of the load. The reliability of the OLTC is directly related to the stability of the transformer and the power system. According to statistics from foreign countries, OLTC faults account for 41% of faults in load-regulating transformers[5][7]. Therefore, the online fault diagnosis of OLTC is a problem that needs to be solved urgently.

The vibration signal is generated during the shifting process of OLTC. When OLTC has hidden troubles or faults, its vibration signal will change. By extracting certain feature from the vibration signal, it can diagnose the status of OLTC, timely repair and improve the reliability of the equipment. Scholars Rivas et al. used hilbert transform and discrete wavelet transform to analyze the vibration signals of different parts of OLTC, and obtained the peak time and amplitude information of the signal envelope[1]. Scholars Zhao Tong and others reconstructed the vibration signal in high-dimensional space, and defined the phase point spatial distribution coefficient to identify the normal and fault states of OLTC

* Corresponding author: peter81015@hqu.edu.cn

[7][8] .Ma Hongzhong and Zhang Huifeng of Hohai University proposed the diagnosis using EMD-HT time-frequency analysis algorithm, EMD entropy and wavelet entropy[2][3].These methods solve the OLTC feature extraction problem to a certain extent, but many methods are limited to feature extraction for a few hundred milliseconds of the in-position section, ignoring the feature information contained in the stationary section.

In this paper, the entire vibration signal of OLTC shifting process is obtained. After the signal pre-processing, it is divided into a stationary section and an in-position section and the energy spectrum of two sections are calculated using the EMD method. The analysis of these energy spectrum samples shows that these method can better identify some failures, such as mechanism jam.

2. EMD energy spectrum algorithm

2.1. EMD energy spectrum algorithm

The EMD method assumes that any signal can be decomposed into multiple IMF components and residuals as shown in Equation 1.

(1) Perform EMD method for vibration signal data $s(t)$. Each IMF component can be regarded as the component of the original signal in a certain frequency band.

$$s(t) = \sum_{i=1}^{n} c_i(t) + r_n(t) \tag{1}$$

$r_n(\mathrm{t})$ is the residual function, which represents the average trend of the signal.

(2) Define the IMF energy as follow, and calculate the energy of each IMF :

$$E(i) = \frac{1}{N}\sum_{i=1}^{N} \left|c_i(t)\right|^2 \tag{2}$$

Where, N is the length of the IMF. The total signal energy is:

$$E = \sum_{i=1}^{M} E(i) \tag{3}$$

Take $\varepsilon_k(i) = E_k / E$, then $\sum_k \varepsilon_k = 1$ .Define EMD energy entropy as:

$$H_k = -\sum_{i=1}^{N} \varepsilon_k(i) \log\left|\varepsilon_k(i)\right| \tag{4}$$

(3) Construct the EMD energy spectrum

Based on the EMD energy entropy by Equation (4), let $H = \sqrt{\sum \left|H_k\right|^2}$ and normalize the EMD energy spectrum of the stationary section.

 https://doi.org/10.1051/e3sconf/202015501006

$$T_1^{'} = \left[\frac{H_{11}}{H_1}, \frac{H_{12}}{H_1}, \cdots, \frac{H_{1k}}{H_1} \right] \tag{5}$$

(4) According to the method of steps (1)-(3), the EMD energy spectrum of the stationary section $T_1^{'}$ and the energy spectrum of the in-position section $T_2^{'}$ are obtained. Therefore, the EMD energy spectrum of the OLTC shifting process T can be defined as:

$$T = \left[T_1^{'}, T_2^{'}\right] = \left[\frac{H_{11}}{H_1}, \frac{H_{12}}{H_1}, \cdots, \frac{H_{1k}}{H_1}, \frac{H_{21}}{H_2}, \frac{H_{22}}{H_2}, \cdots, \frac{H_{2k}}{H_2} \right] \tag{6}$$

2.2. Calculate correlation of the EMD energy spectrum

Considering that the energy spectrum has been normalized, the Pearson correlation coefficient is used to determine the correlation between different signal samples. The Pearson correlation coefficient is defined as:

$$\rho_{x,y} = \frac{E((x-u_x)(y-u_y))}{\sigma_x \sigma_y} = \frac{E(xy)-E(x)E(y)}{\sqrt{E(x^2)-E^2(x)}\sqrt{E(y^2)-E^2(y)}} \tag{7}$$

Where x and y are the energy spectrum of the samples, $E(x)$ and $E(y)$ are the mean of them. If the feature extraction method is effective, the correlation in the same state will high and the correlation in different states will low. For this article, when $\rho > 0.95$, the samples are considered to be similar, and when $\rho < 0.85$, the samples are considered different.

3. OLTC vibration signal preprocessing

The accelerometer sensor is installed on the head end cap of OLTC ZVIII350. The vibration signal of OLCT collected by the accelerometer is sampled by the data collector and uploaded to the data processing software on the PC in the experiment.

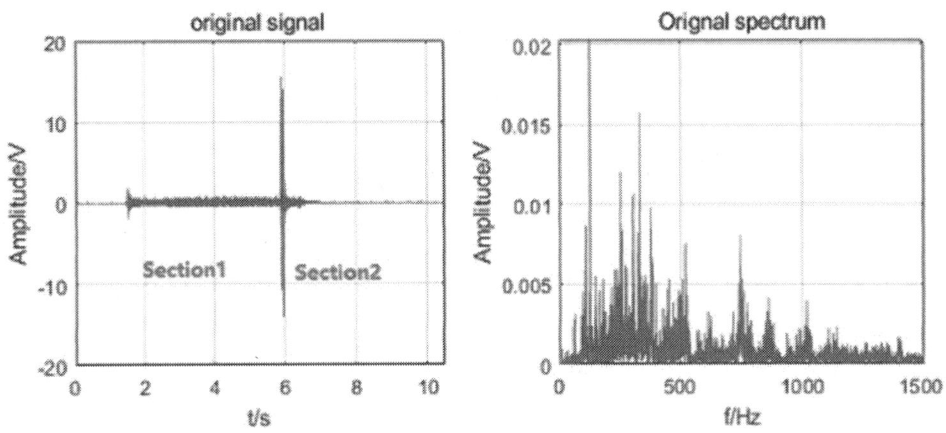

Figure 1. Original vibration waveform and signal spectrum

From the related literature [6] and the signal spectrum in the right part of Figure 1, it can be found that the energy is mainly concentrated at 50Hz-1kHz. Combining the principle of OLTC and the characteristics of the vibration signal, the entire shifting process can be divided into two section: the stationary section and the in-position section. The stable section is mainly the external electric mechanism driving the spring to store energy. At this stage, the vibration signal is relatively stable and the amplitude is small. The in-position section is the release of the over-dead point of the energy storage mechanism, the switching component operates according to predetermined requirements, and the contacts are switched on and off to achieve gear switching, and the vibration signal changes suddenly.

Taking the signal samples in the normal state as an example, at this time, the OLTC is increased from 1st gear to 2nd gear. The vibration signals of the stationary section and the in-place section are solved by empirical mode decomposition to obtain IMF component, and the odd components are shown in the figure below.

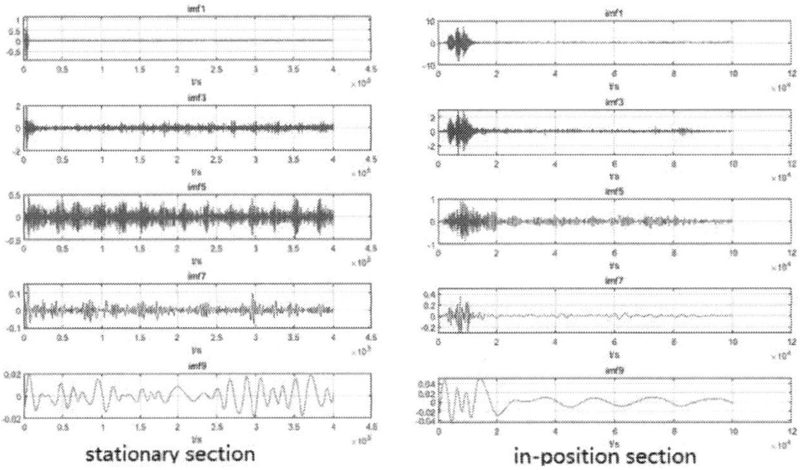

Figure 2. EMD diagram for two section

As shown by the IMF9 waveform in Figure 2, the components IMF8-IMF10 have obvious modal aliasing, and their signal amplitude are only one tenth of other components. Therefore, the characteristic energy spectrum of OLTC is composed of the first 7 IMF components of the stationary section and the in-place section, that is, a total of 14 energy Energy entropy form the EMD characteristic energy spectrum.

4. Energy spectrum extraction and analysis

The OLTC vibration experiment obtained a total of 192 sets of vibration signal samples, of which 97 groups were in normal state, 50 groups were mechanism jam, and 45 groups were spring breakage. Five signal samples are taken for each state, and the Pearson correlation coefficient is used for data analysis.

 https://doi.org/10.1051/e3sconf/202015501006

4.1. EMD energy spectrum in normal state

The energy spectrum of five samples in the normal state is shown in the figure below.

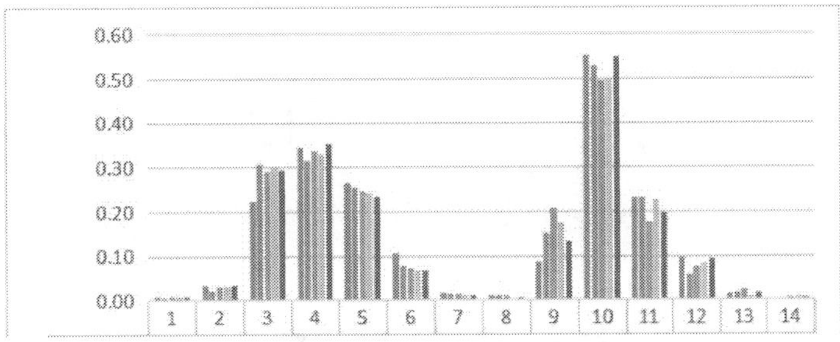

Figure 3. EMD energy spectrum in normal state

It can be seen that the distribution characteristics of the energy spectrum of vibration signal are consistent. The energy is mainly concentrated in the IMF3≈0.29,IMF4≈0.32, IMF5≈0.24 in the stationary section, and the IMF9 ≈0.15, IMF10≈0.52, IMF11≈0.2 and IMF12≈0.1, etc in in-position section. The table 1 shows the correlation of the five samples in normal state is greater than 0.95, and the similarity of the samples is high.

Table 1 The correlation of EMD energy spectrum in normal state

Correlation	Normal 1	Normal 2	Normal 3	Normal 4	Normal 5
Normal 1	1.000	0.978	0.985	0.960	0.972
Normal 2	0.978	1.000	0.992	0.988	0.996
Normal 3	0.985	0.992	1.000	0.987	0.992
Normal 4	0.960	0.988	0.987	1.000	0.994
Normal 5	0.972	0.996	0.992	0.994	1.000

4.2. EMD energy spectrum of mechanism jam

The energy spectrum of signal samples under jam conditions is shown below.

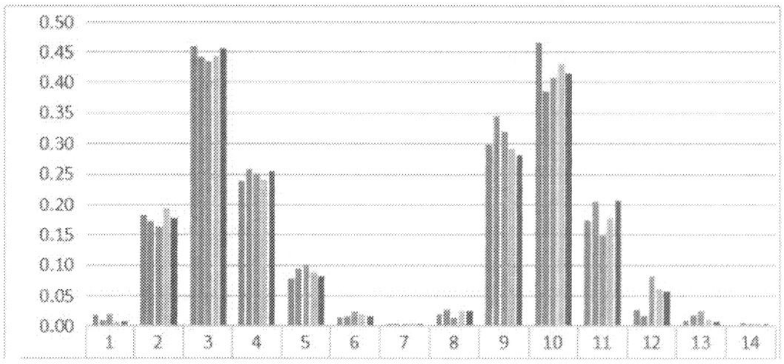

Figure 4. EMD energy spectrum of mechanism jam

The figure shows that the characteristics of the energy spectrum of the mechanism jam faults have good consistency. The corresponding frequency of the IMF component is from high to low. Compared with the normal state, the energy of the high-frequency vibration signal is increased. The IMF2 ratio of the stationary section has increased significantly to 0.17, the IMF3 ratio has increased to 0.45, and IMF4 has decreased slightly to 0.25. The IMF9 ratio of the in-position section increased to 0.3 and IMF10 decreased to 0.4.The correlation of the characteristic energy spectrum of samples is also above 0.95.

Compared with the same state, the correlation between different states are less than 0.85 in Table 2 below, which indicates that it can distinguish between normal state and mechanism jam.

Table 2 The energy spectrum correlation between mechanism jam and normal state

Correlation	Jam1	Jam 2	Jam 3	Jam 4	Jam 5
Normal 1	0.727	0.684	0.715	0.725	0.726
Normal 2	0.828	0.798	0.818	0.824	0.827
Normal 3	0.818	0.779	0.882	0.816	0.816
Normal 4	0.818	0.826	0.852	0.844	0.842
Normal 5	0.843	0.821	0.843	0.843	0.847

4.3. The effectiveness of the stationary energy spectrum

In order to verify the effectiveness of the method, we take the normal state and the mechanism jam as examples, as shown in the Table 3.

Table 3.Comparison of correlation before and after adding stationary features

	without the stationary section					with the stationary section				
Correlation	Jam 1	Jam 2	Jam 3	Jam 4	Jam 5	Jam 1	Jam 2	Jam 3	Jam 4	Jam 5
Normal 1	0.862	0.757	0.805	0.860	0.867	0.727	0.684	0.715	0.725	0.726
Normal 2	0.928	0.849	0.876	0.924	0.930	0.828	0.798	0.818	0.824	0.827
Normal 3	0.907	0.809	0.942	0.905	0.905	0.818	0.779	0.882	0.816	0.816
Normal 4	0.907	0.900	0.941	0.968	0.964	0.818	0.826	0.852	0.844	0.842
Normal 5	0.942	0.872	0.904	0.944	0.949	0.843	0.821	0.843	0.843	0.847

It can be seen that the correlation between different states is obviously reduced after the characteristic energy spectrum of the stationary section is supplemented,which indicates that the characteristics energy spectrum of the stationary section is correct and effective.

It should be noted that some states such as contact failure only occur in the in-position section, and the characteristics of the stationary section are not affected. In this case, there will also be a situation where the correlation is slightly increased after adding the characteristics of the stationary section.

5. Conclusion

Aiming at the problem that the original OLTC fault diagnosis only focused on the vibration signal of the in-position section, this paper proposes to increase the characteristic energy spectrum of the stationary section as a supplement, and together with the in-position section's characteristic, form the EMD energy spectrum. Without increasing the number of sensors and the complexity of the algorithm, the effectiveness of state feature extraction is improved. The research in this paper shows that the EMD energy spectrum in the same state has regularity and high similarity, but the EMD energy spectrum in different states has obvious differences.

Funding

The authors disclosed receipt of the following financial support for the research, authorship, and/or publication of this article: Xiamen Science and Technology Plan Project

(Project No.3502Z20193032), and Natural Science Foundation of Fujian Province, China (Project No. 2019J01054).

References

1. Rivas-Trujillo E , Edwin Laniado-Jácome, José Luis San-Román, et al. Characterizing the diverter switch of a load tap changer in a transformer using wavelet and modal analysis[J]. Engineering Structures, 2010, **32**(10):3011-3017.
2. Zhang Huifeng, Ma Hongzhong, Chen Kai, et al. Fault diagnosis of transformer on-load tap-changer based on time-frequency analysis of vibration signal emd-ht [j]. High Voltage Apparatus, 2012, **48** (1): 76-81.
3. Gao Peng, Ma Hongzhong, Zhang Huifeng, et al. Comparison of emd entropy and wavelet entropy of tap-changer vibration signal [j]. Journal of Electric Power Systems and Automation, 2012, **24** (4): 48-53.
4. Wang Chunliang, Li Kai, Ma Hongzhong, et al. Feature recognition of oltc vibration signals based on wp_svd noise reduction [j]. Journal of Electric Power Systems and Automation, 2012, **24** (5): 36-41.
5. Wang Mengyun. Statistical Analysis of 110kV and Above Transformer Accidents in 2004 [J]. Electric Power Equipment, 2005, **6** (11): 31-34..
6. Hong Xiang, Ma Hongzhong, Gao Peng, et al. Fault diagnosis of loose contact of on-load tap-changer based on EEMD [J]. HUA DIAN TECHNOLOGY, 2012, 34 (1): 12-15.
7. Zhao Tong. Research on on-line monitoring and fault diagnosis of on-load tap-changer mechanical status [d]. Shandong University, 2008.
8. Zhao Tong, Li Qingmin, Chen Ping. Dynamic analysis method for feature extraction of oltc vibration signals [j]. Journal of Electrical Engineering and Technology, 2007, **22** (1): 41-46.

A Review of Research Status and Prospect of Vacuum Insulated Tubing Insulation System

*Xing Yu**, *Liu Lei, Cao Chang, Fan Weidong, Yan Keju,Cheng Zhongfu*

Sinopec Northwest Oil Field Company,Wulumuqi 830011, China

Abstract. Vacuum heat insulating tubing is an important wellbore heat insulating facility for heavy oil thermal recovery at present. Its heat insulating performance directly affects the thermal recovery efficiency. The research on the heat insulating system of vacuum heat insulating tubing is of great significance to improve and improve the heat insulating performance and enhance the thermal recovery of heavy oil. This paper summarizes and analyses the current research status of vacuum insulated tubing insulation system. It elaborates the insulation structure, insulation materials, annular air charging and vacuum pumping, insulation coating and so on. It provides reference and reference for the future research of vacuum insulated tubing.

Keywords :vacuum insulated tubing; thermal insulation; review

1 Introduction

Insulated tubing is a professional string for thermal recovery of heavy oil, and is one of the necessary downhole tools for heavy oil thermal recovery. Steam injection through insulated tubing can bring more heat into the reservoir and improve the dryness of steam reaching the bottom of the well, thus gaining the effect of heavy oil development and production increase [1]. It is of great significance to carry out the research on steam flooding heat insulation tubing for promoting the development of steam flooding heavy oil thermal recovery technology and the long-term development of enterprises [2, 3]. Vacuum insulating tubing is the most effective insulating technology in the world, so it is also called "super insulating technology" (VIT). On the basis of investigating the domestic and foreign literature on heat-insulating tubing, this paper analyses and summarizes the research on heat-insulating structure system and material of heat-insulating tubing, and looks forward to the future research prospects.

* Corresponding author： Li964499@126.com

2 Domestic vacuum insulation tubing research status

At present, the main manufacturers of heat insulation oil pipes in China are Liaohe Petroleum Exploration Bureau General Machinery Factory, Shengli Oilfield Furuite Petroleum Equipment Co., Ltd., Jiangsu Yancheng Jinfeng Petroleum Machinery Company and Wuxi Simles Steel Pipe Co., Ltd. Liaohe Petroleum Exploration Bureau General machinery plant is the largest manufacturer of vacuum insulation tubes in China [4].

In the study of the effect of vacuum on the thermal insulation performance of vacuum insulated tubing in the General Machinery Factory of Liaohe Petroleum Exploration Bureau, it is pointed out that pumping the internal pressure of the insulating layer into (30-40) Pa compartments is very helpful to ensure the thermal insulation performance of the insulating tubing and control the production cost of the insulating tubing[5].

In the study of improving the wellbore thermal insulation and cold production technology of low-cost insulated tubing in Shenyang Oil Production Plant, the strength of prestressing force and high temperature treatment on production tubing string is reduced. The vacuum degree is within an optimal range. The improved vacuum insulated tubing has been in well time for half a year, and its apparent thermal conductivity is still in grade D, basically stable and applied. The service life can be increased to 3.5-4 years. In order to ensure the vacuum life of the insulating sleeve, special high temperature air-breathing materials are placed in the annulus. The technology of low cost vacuum insulated tubing is applied in 38 wells, and the total electricity saving is 542.68×104kW. h[6].

In the research and application of high condensate oil cold production technology in Shenyang Oilfield, the selected target well is 67-563, which is produced by 1200m vacuum heat insulation tubing. For oil wells with 18 tons of liquid per day, the wellhead temperature can be increased by 10-13 ℃, and the outlet temperature can reach 37 ℃, which can achieve the purpose of cold production; 1600m vacuum heat insulation tubing can be used to maintain the wellhead temperature. Temperature production can increase wellhead temperature by 12-15 ℃ for oil wells with daily liquid production of more than 16 t, and due to the temperature of wellhead liquid production reach about 37 ℃, so as to achieve the goal of cold production [7].

In the numerical simulation analysis of the thermal insulation effect of foam glass insulated tubing and vacuum heat insulated tubing, the Chinese Academy of Sciences and CNOOC Research Institute have found that before the hydrogen permeation, both the vacuum insulated tubing and the foam glass heat insulation oil pipe can achieve good thermal insulation effect. However, after the hydrogen permeation, the heat flux density of the vacuum insulated tubing is urgent. The increase of the theater result in the failure of the vacuum insulated tubing and the increase of heat flux density after the hydrogen permeation of the foam glass insulating oil pipe is not obvious. When the thickness of the insulation layer of the foam glass heat insulation pipe reaches 16mm, the heat flux density of the foam glass insulation tube is actually equivalent to that of the vacuum insulation tube before failure[8].

The Research Institute of China National Offshore Oil Corporation has put forward a heat insulation technology which combines thermal insulation coating with vacuum technology[9]. The main design method of this process is: heat insulation coating treatment of the outer wall of the insulating oil pipe, high temperature and high pressure packer is placed in the designated position of the casing annulus, vacuum pump is used to vacuum the casing annulus through the casing wing valve of the oil production tree, and heat injection operation can be started after meeting the requirements (vacuum is installed on the casing wing valve). The vacuum degree of oil sleeve annulus can be monitored during the injection heat, and the vacuum can be supplemented according to the requirement.

Shengli Oilfield has developed the fourth and fifth generation heat insulating pipes, hydrogen-proof heat insulating pipes and vacuum heat insulating pipes, which have entered the field use and test stages respectively [10]. Under the condition of steam injection at 325 ℃, the apparent thermal conductivity of the vacuum insulating pipe is still 0.008W/(m.K) after three cycles of test. The anti-hydrogen damage insulating pipe produced by new hydrogen absorbent can double the service life of the original insulating pipe and reach 30 huff and puff cycles. Steam injection string has been able to meet the needs of different heat insulation requirements for 200-1700 m wells, especially for 800-1700m deep wells. The wellbore heat loss can be controlled within 12% by using high efficiency heat insulation pipe, high temperature separator, expansion pipe and annulus filled with nitrogen gas. The heat insulation performance has exceeded that of the products imported from abroad in the late 1980s.

3 Foreign research on vacuum insulated tubing

At present, the technology of high-efficiency heat preservation and insulation abroad is relatively mature, and the technology of vacuum heat insulation tubing is widely used in the world. The application of VIT in Troika Oilfield in Britain keeps the fluid in the wellbore at a higher temperature and reduces energy consumption; the high-efficiency heat insulation pipe used in Kern River Oilfield in the United States reduces the energy loss of the wellbore by about 5%; the heat insulation tubing has been applied in actual production in Gulf of Mexico [11] and offshore oil field in Brazil [12].

The VIT produced by ConocoPhillips is successfully applied to in waxy crude oil extraction technology under arctic environment [6]. In the wellbore of deep water oil field in Gulf of Mexico, VIT has been successfully used for heat insulation, solving the problems of paraffin deposition and hydrate blockage, which hinder the flow of well fluid. In the tubing of Norman well, VIT can greatly reduce the radial heat transfer of wellbore and solve the problem of paraffin deposition in inner tubing.

VIT has been widely used in steam injection wells in onshore oilfields in northeastern and southeastern Brazil [13]. It has also been tried in offshore well production. The structure sketch and details of connection are shown in Fig. 1.

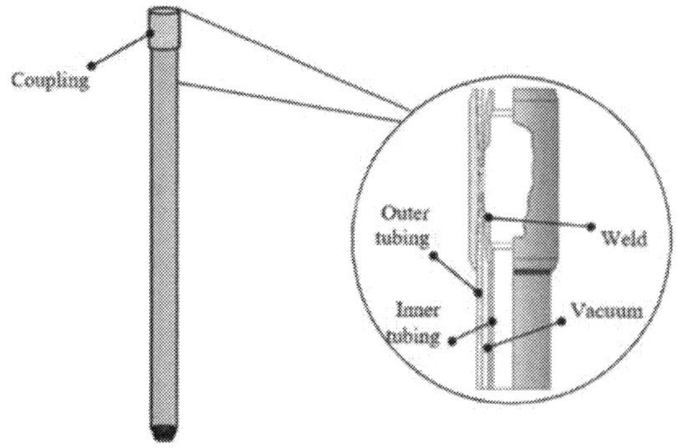

Fig.1. Vacuum insulated tubing concept and detail of the connection region.

Exceed High Thermal Efficiency Tubing (HTET) manufactured by Exceed oilfield equipment Company of Canada uses new nano-insulation materials, which can reduce steam heat loss, maintain steam quality, reduce casing deformation and protect cement casing. Canada also has a new type of heat insulating tubing called T-IMEX. Unlike the traditional VIT, IMEX has only one leakage path, which further reduces heat loss. The structure and details of HTET are shown in Figure 2.

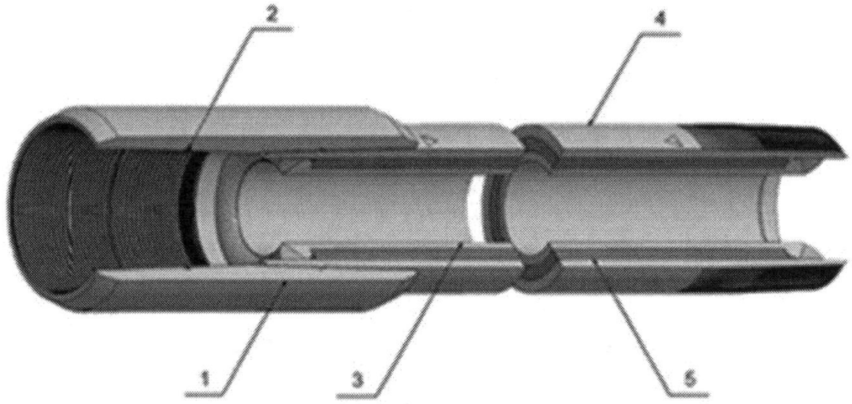

1. Coupling with USS Buttress (Coupling for Outer Tube Threads)
2. Seal Ring
3. Inner Tube-API 5CT gr N80 range 2
4. Outer Tube-API 5CT gr N80 range 2
5. Thermal Insulation

Fig.2. Vacuum insulated tubing concept and detail of the connection region.

T Śliwa [14] studied the application of vacuum insulated tubing in deep well heat exchangers. The biggest problem in the production and development of VIT is hydrogen permeation. This process affects the insulation efficiency and increases the heat transfer through the inner column. Hydrogen is easily penetrated into most steel materials. Absorbent is a good absorbent to absorb hydrogen through the vacuum ring. Therefore, before installation of VIT, the amount of absorbent material and potential hydrogen permeation need to be studied through field test or laboratory research.

In 2015, at the SPE Hot Well Integrity and Design Symposium held in Banff Alberta, Canada, J. Damour and other scholars introduced their work on steam flow in a project in the Middle East. The design and economic considerations of selecting VIT or VIC in hot well completion are introduced. J. Damour summarized the main failure modes of oil well integrity: casing expansion and the resulting load, cement integrity under thermal cycling, annulus pressure accumulation. In the BP MARLIN project, production casing collapsed due to annular pressure, resulting in well loss [15]. The drill string of Pomp.A-31 well in the Gulf of Mexico was distorted by annular pressure, which caused 16 inch casing stuck [16]. At the same time, the seabed wellhead prevented deep water wells from releasing annular pressure in casing annulus, which is different from the onshore wells and platform wells. In addition, deepwater drilling accidents will not only cause huge economic losses, but also cause serious environmental pollution [17]. Therefore, scholars have provided a solution for evaluating VIT (or VIC) by thermal well models of casing and tubing temperatures in order to avoid well failure. The heat loss of different VIT products is given to measure the impact on the process efficiency and economy. Some parameters considered in the design of hot well are given: casing temperature, tubing temperature and heat loss. By determining these parameters, the integrity of the well can be studied, and the efficiency of the whole process is estimated to be [18]. They quantitatively analyzed the influence of thermal performance and tubing size on underground steam quality. The steam flow analysis model was used to compare the different diameter and heat insulation performance of the same well. The thermal performance (total heat transfer coefficient) of different types of heat insulation pipes is determined by the operator through field tests before being used as model input. The output of the model is the mass, temperature and pressure of the steam, which is a function of the well location (including downhole conditions). It is also found that under the optimal thermal performance, reducing the diameter of the pipe can have a positive impact on the economy of the system. On the contrary, it is found that it is impractical to reduce the size of pipes by using pipes with relatively poor thermal performance. The results can be used to select the type and size of the new syringe Wells tube. With the completion of the analysis, steam wells are being modified according to research proposals in the field of research. Vacuum heat insulation pipes are gradually replaced by design with improved thermal performance, and the size of the pipes is reduced, thus reducing the overall cost and improving the quality of downhole steam [19].

Rashid Al Shaibi (SPE 172930-MS) describes an experiment carried out by operators of the Middle East Heavy Oilfield in steam injection wells, in which heat insulation pipes from four different manufacturers are installed, in which temperature can be monitored by optical fibers and thermometers. The data collected during the experiment show that the vacuum insulation technology has the best effect on the joint: the overall thermal performance is the best, the thermal performance will not deteriorate over time, and the temperature difference between the joint and the joint body is the smallest. The three combinations that best represent the different solutions of oil wells are: (1) insulating pipes with 3 103 1/2 inner pipes, which have standard thermal performance; (2) insulating pipes with 3 103 1/2 inner pipes, which have excellent heat dissipation performance; (3)

insulating pipes with 2 103 7/8 inner pipes, which have excellent heat dissipation performance. For each configuration, steam is injected at 572 degrees F (300 degrees C) with a mass of 85% at the wellhead[19].

Mirko Zatka of Shell Energy Canada presented a special report entitled "Application of Vacuum Thermal Insulation Pipe in Hot Oil Sand Engineering" at the SPE Heavy Oil Conference and Exhibition held in Kuwait City in 2016. The report traces the technical development history of VIT, introduces the manufacturing process of VIT and design elements affecting its long-term performance. How to determine the actual heat transfer coefficient of the product, and how to verify the real performance once the VIT is installed in the well, et. al. [20].

ITM and OTSI use proprietary software programs to calculate and predict bottomhole steam quality. Field tests by industry and the U.S. Department of Energy confirm the validity of software programs used to predict heat loss and bottomhole quality [21]. The test results show that compared with bare tube, vacuum insulation tube will greatly reduce heat loss and improve bottom hole steam quality. At 1,000 feet (304.8m), the difference between insulated and non-insulated tubing is 13.6%, increasing to 47.5% at 3,500 feet (1066.8m). Checking the percentage of steam mass in each type of tubing, the researchers found that the steam mass of bare tubing decreased by 35.4% from 1000 to 3500 feet (305-1067m), with the steam quality of insulated tubing almost unchanged.

Editor of Journal of Petroleum Technology (JPT) pointed out that the most common methods of thermal enhanced oil recovery are cyclic steam agitation, steam flooding and steam-assisted gravity drainage, which are widely used in Canada [22]. Typical steam injection temperatures are between 250 and 310 degrees Celsius, and may be greater than 310 degrees in a few cases. This high temperature application requires adjusting the design of injection wells to avoid any mechanical failure of wells or to reduce heat loss through wells. Vacuum insulation tube (VIT) or vacuum heat insulation casing (VIC) is a way to solve these problems. There are two main types of insulated tubing. The first type of VIT is based on the thermal insulation materials used in the aerospace industry. It needs high vacuum (10-5 Torr) to connect to multi-layer foil. The hydrogen leakage of carbon steel through pipeline increases the vacuum pressure in annular space. Therefore, the use of high vacuum leads to the decrease of the service life of VIT. Another kind of thermal insulation material is the microporous material used between the inner and outer pipes. The application of soft vacuum, usually 10 torus, reduces the heat transfer mode by reducing the average free path of the air molecules in the microporous material.

In comparing concentric and eccentric configurations of VIT and bare tubing, ANSYS Canada and ZEroCor Pipeline Company studied three different formation temperatures (253.15K, 313.15K and 423.15K), two different tubing types (bare tubing and VIT tubing) and two different tubing configurations (concentric tubing and eccentric tubing). 12 simulations of were carried out [23]. On the other hand, the steam quality of VIT tubing is minimized at different formation temperatures. The vacuum insulation material of VIT can be used as a good insulation layer and keep heat in the pipeline. Under three different temperatures, the variation of steam mass is less than 1%. Compared with bare tubes, the steam quality is saved by 4.6% on average.

https://doi.org/10.1051/e3sconf/202015501007

4 Conclusions and prospects

This paper summarizes and analyses the research on the thermal insulation structure system and material of vacuum insulated tubing at present, and summarizes the current research and development status. It can be seen from the research at home and abroad that vacuum insulated tubing is a widely used wellbore thermal insulation technology at present. Its product quality is reliable and its performance is stable, which makes its tubing body. The apparent thermal conductivity is reduced by one order of magnitude than the apparent thermal conductivity of the conventional insulating tubing. However, the heat flux density of vacuum insulating pipe after hydrogen permeation failure is very large, which is more than 10 times that before failure. In order to maintain its excellent heat insulation performance in the long-term steam injection process and achieve excellent heat insulation effect, it is necessary to inject getter in annulus, absorb H2 and other residual gases infiltrated into steel pipe during steam injection process, so as to make the heat insulation system. The system will be purified at any time to maintain the intact state of the vacuum system. From the point of view of technology and economy, vacuum insulated tubing has high price, complex structure and self-weight, which is not suitable for deep well operation. The research and development of new thermal insulation materials were predicted. It has remarkable effects on thermal insulation and thermal recovery, and is also the focus of future research.

References

[1] Bi Pu-yue, Liu Xin-tian. Application and analysis of thermal insulation technology for hot oil injection steam pipe network [J]. Chemical Engineering Design Communications, 2017, 43(12):61-61.

[2] Guo Qian. The research and application of composite insulation lining tubing technology[D]. Southwest Petroleum University, 2017:1-5.

[3] Zhang Xiao–hui. Preparation and Property of Thermal Insulation Coating on the Connection Between the Insulated Tubing[J]. materials protection 2018, 51(2): 82-84.

[4] Sun Cheng-li. Development and application of insulated tubing used for heavy oil thermal production[D]. Northeast Petroleum University, 2014:2-6.

[5] Wu Yong-ning. The Vacuum Degree Impact on the Insulation Properties of Insulated Tubing[J]. Science Technology and Engineering, 2010, 10 (11): 2728 - 2732.

[6] Zhao Xu-guang, Wang Zhan-sheng. Research and application of low temperature heat insulation tubing shaft insulation technology[J]. Chemical Engineering Design Communications,2017, 43(8):76-76.

[7] Guo Dao-hong. The study and application of nonthermal production technology of high pour point crude oil in Shenyang Oil Field[D]. Northeast Petroleum University, 2012:27-30.

[8] Zhou Cheng-long, Xu Yong-xiang, Sheng Hong-zhi, et. al. Numerical simulation analysis of thermal insulation effect of foam glass insulated tubing and vacuum insulated tubing[J]. Chemical Engineering & Equipment, 2015(1):5-9.

[9] Lin Tao, Zhang Wei, Zou Jian, et.al. Heat insulation coating and vacuum technology and its application in heavy oil thermal production[J]. Petroleum Geology and Engineering, 2016(5):127-129.
[10] Liu Jun-yan.Reesareh on improving steam thermal efficiency[D]. Southwest Petroleum University, 2003:35-44.
[11] S.W. Gosch, D.J. Horne, P.D. Pattillo, J.W. Sharp, P.C. Shah, Marlin failure analysis and redesign: part 3 – VIT completion with real-time monitoring, SPE Drill.Complet. 19 (02) (2004) 120–128.
[12] M.V.D. Ferreira, A.R. Dos Santos, V. Vanzan, Thermally insulated tubing application to prevent annular pressure buildup in Brazil offshore fields, SPE Deepwater Drilling and Completions Conference, Galveston, Texas, USA, June 20–21, 2012.
[13] Barcelos J G A, Ferreira M V D, Costa C A S, et al. Thermal analysis of vacuum insulated tubing (VIT) for offshore wells[C]// Brazilian Congress of Thermal Sciences and Engineering. 2014.
[14] Śliwa T, Kruszewski M, Sapińska-Śliwa A, et al. The application of Vacuum Insulated Tubing in Deep Borehole Heat Exchangers[J]. 2017, 34(2):597.
[15] D.W. Bradford, D.H. Gibson, S.W. Gosch, P.D. Pattillo, J.W. Sharp, C.E. Taylor,Marlin failure analysis and redesign: part 1 description of failure, IADC/SPE Drilling Conference, Dallas, USA, February 26–28, 2002.
[16] P.D. Pattillo, B.W. Cocales, S.C. Morey, Analysis of an annular pressure buildup failure during drill ahead, SPE Drill. Complet. 21 (04) (2006) 242–247.
[17] L.-G. Faksness, D. Altin, T. Nordtug, P.S. Daling, B.H. Hansen, Chemical comparison and acute toxicity of water accommodated fraction (WAF) of source and field collected Macondo oils from the Deepwater Horizon spill, Mar. Pollut. Bull. 9(01) (2015) 222–229.
[18] Damour J, Johannson D. What Vacuum Insulated Tubings Or Vacuum Insulated Casings Bring To Thermal Wells[C]// SPE Thermal Well Integrity and Design Symposium. 2016.
[19] Damour J, Brown T. Downsizing Steam Injection Insulated Tubing through the Use of Superior Thermal Performance[C]// SPE Thermal Well Integrity and Design Symposium. 2015.
[20] Zatka M. Application of Vacuum-Insulated Tubing VIT in Thermal Oil Sand Projects[C]// SPE Heavy Oil Conference and Exhibition. 2016.
[21]Lombard, M. S., Lee, R., Manini, P., & Slusher, M. S. New Advances and a Historical Review of Insulated Steam-Injection Tubing. [C]//SPE Western Regional and Pacific Section AAPG Joint Meeting. 2008.
[22]Carpenter, C. What Vacuum-Insulated Tubings and Casings Bring to Thermal Wells[J]. Journal of Petroleum Technology, 2017,69(06), 73–74.
[23]Karra, S., Chudiak, M., Sinha, A., & Boser, G. SAGD Vacuum Insulated Tubing vs. Bare Tubing - Concentric and Eccentric Configurations: A Comparative Thermal Computational Fluid Dynamics Study. SPE Heavy Oil Conference-Canada. 2014.

Analysis of economic and energy consumption driving factors of China's construction industry investment in 2015

Zijian Zou[1], *Bo* Bai [2,*]

[1,2]School of Environment and Energy, Peking University , Nanshan District, Shenzhen, China

Abstract. This paper uses the input-output model to analyse the economic and energy consumption impact of China's construction investment in 2015, and finds that it can drive economic growth of 15.11 trillion in total, with direct and indirect impacts of 27.22% and 72.78% respectively , and investment can actively promote the development of integrated services and finance and insurance sectors. Additionally, the direct energy consumption caused by the investment is 20.98Mtce, accounting for only 1.23%, while the indirect energy consumption is 1701.10Mtce, accounting for 98.77%, of which the largest energy consumption sector is the industry of Production and supply of electricity and heat, the industry of Metal smelting and rolling processing, the industry of non-metallic mineral products, indicating that the construction industry development of china has a strong dependence on traditional energy sources. In the future, we are supposed to focus on indirect energy consumption in the construction industry and formulate morae effective clean energy-saving policies.

1 Introduction

In economic theory, investment, consumption, and export are often compared to the troika of economic growth. Many scholars believe that there is a strong positive correlation between economic development and investment, and especially the impact of productive capital investment on economic growth.(Chaudhri,2010; Giray Gozgor,2018; Qianxue Zhang,2018; Tahsin Bakirtas,2018) A large amount of investment has effectively promoted economic development in china, and fixed asset investment indicators are often used as a barometer of China's industrial development.

Studies by scholars such as Feng Qian and others (2012) show that the impact of China's fixed asset investment on CO2 emissions is significantly positive. With the emergence of climate change and its questioning of human beings' pure pursuit of economic growth, curbing the growth of carbon emissions and sustainable growth have become new goals for human beings.

The Harvard China Carbon Emissions Report 2015 shows: "Industrial production and thermal power generation are the most important carbon emission sectors in China, and

* Corresponding author: bobai@pku.edu.cn

they together contribute 85% of carbon emissions in china. The manufacturing and construction industries aggregate is the most significant reason for high carbon emissions in china. " However, with the deterioration of the environment and the strengthening of energy constraints, the growth of energy consumption caused by China 's fixed asset investment has become more serious. Reducing carbon emissions caused by Chinese investment is crucial for China to complete carbon reductions and build a sustainable economic development system.

China's main fixed asset investment is concentrated in the construction industry. According to the statistics of the fourth assessment report of the IPCC and the "Building Energy Consumption Research Report" released in 2016, it is known that the mean energy consumption of the global construction industry accounts for 40%. The average carbon emissions reached 36%. In 2015, the energy consumption consumed by the investment in construction accounted for nearly one-third of China. The energy consumption of construction accounted for a relatively high proportion. Therefore, China's current reduction target for total emission reduction should focus on the construction industry, and the specific effect will directly affect the achievement of China's 2030 emission reduction target.

Many researchers have studied the affect of construction industry investment on economic growth and energy consumption from different perspectives. Hui Yan (2012) and others divided the construction process into four departments: manufacturing, transportation, installation, and construction of building materials. Taking Hong Kong as an example, the analysis showed that these four processes all generate greenhouse gas emissions, with the consumption of building materials producing the most greenhouse gases. Sandanayake (2015) and others carried out a hierarchical algorithm on the construction process to analyze the carbon emissions generated in the building construction. They conducted a directional analysis of the two buildings and found that the use of equipment, The carbon emissions from transportation and the use of building materials are: 19%, 14%, 67%. The researches of the above scholars have paid close attention on all the energy consumption generated by the investment in the construction, but the construction industry is a typical "low-carbon, high-implicit carbon" industry. Generally, investment in the construction industry not only causes direct carbon emissions, Usually, it also affects the emissions of other sectors by providing products to other sectors. For example, the material consumption caused by the development of the construction industry will generate a lot of indirect economic growth and indirect carbon emissions. Therefore, this article will focus on the direct and indirect economic growth and energy consumption generated by the construction investment in 2015, comprehensively assess the economic and energy consumption impact of China's construction industry investment, and provide policy makers with more specific policies suggest.

Input-output analysis is an effective way to analyze the economic growth and industrial structure change of an economic system driven by demand factors such as investment and consuming.Liu (2017) and others used the IOA to assess the contribution of China's industrial SO2 and COD emission factors and found that the reduction in energy consumption in 2007-2010 was due to the structural adjustment of the construction sector. Jindao Chen et al. (2017) article found that from 1995 to 2011, the carbon dioxide emissions of construction industry in china increased by 388.7%, and indirect accounted for 12.7%. Indirect energy consumption is not adequately addressed in practice and theory. The studies of the above scholars are concentrated before 2012. Since the carbon emission model of the construction industry will change with the adjustment of the economic structure after 2012, this article uses the input-output table of 2015 to analyze the investment impact of the construction industry. It is of great significance to achieve economic goals and energy conservation.

2 Method and data sources

2.1 Input-output model

The economic activities will always be accompanied by the generation of energy consumption. With the technological conditions and the industrial structure unchanged, the more output is, the more energy will be consumed. The energy consumption associated with economic activities is implied in the products. As an exogenous factor of economic growth, investment not only expands the output of the invested industry directly but also promotes the growth of other industries indirectly through the correlation between industries. Since economic activities inevitably generate energy consumption, the corresponding carbon emissions can be calculated from the carbon content contained in the investment-driven output and the required energy consumption. The total economic output driven by investment can be evaluated from the Leontief model. Assuming there are N departments, the Leontief model is:

$$X = (I - A)^{-1} \times Y = (I + A + A^2 + \cdots) \times Y = L \times Y \quad (1)$$

Where

$$A = \begin{bmatrix} a_{11} \cdots a_{1n} \\ \vdots \quad \vdots \\ a_{n1} \cdots a_{nn} \end{bmatrix}, \quad Y = \begin{bmatrix} y_1 \\ \vdots \\ y_n \end{bmatrix}, \quad X = \begin{bmatrix} X_1 \\ \vdots \\ X_n \end{bmatrix}, \quad L = (I - A)^{-1} = \begin{bmatrix} l_{11} \cdots l_{1n} \\ \vdots \quad \vdots \\ l_{n1} \cdots l_{nn} \end{bmatrix}$$

In the formula (1), **X** means to the total output column vector. **A** means direct consumption coefficient matrix.**I** means identity matrix. **L** means Leontief inverse matrix. **Y** presents final demand column vector of the investment. The elements in the above matrix represent the sectoral matrix, such as $\mathbf{a_{ij}}$ means direct consumption matrix of the *i* Department considering unit input from the *j* department. Take occurred investment of real assets as an input variable into matrix **Y** of formula (1), and the occurred total economic outputs **X** can be calculated.

The energy consumption produced by the unit output is defined as the energy consuming coefficient **F**, and the increment value produced by the unit output is defined as the increment coefficient **G**. The calculation formula is as follows.

$$F_{ij} = \frac{f_{ij}}{X_j} \qquad (i, j = 1, 2, \cdots, n) \quad (2)$$

$$G_{ij} = \frac{g_{ij}}{X_j} \qquad (i, j = 1, 2, \cdots, n) \quad (3)$$

Where $\mathbf{X_j}$ indicates the total output of the *j* department, $\mathbf{f_{ij}}$ means to the total energy consumption for the department *j*, and $\mathbf{g_{ij}}$ means to the increment value for the department *j*.

Then, the increment value of the finance, $\mathbf{X_G}$, which is driven by the investment **Y**, can be calculated by the following formula.

$$X_G = G(I - A)^{-1} \times Y = G \times L \times Y \quad (4)$$

Increment value of the finance,$\mathbf{X_G}$, which is driven by the investment **Y**, can be calculated by the following formula.

$$X_F = F(I - A)^{-1} \times Y = F \times L \times Y \quad (5)$$

Then we introduce the energy consumption coefficient diagonal matrix $\hat{F}$, the increment value coefficient diagonal matrix $\hat{G}$ and the investment diagonal matrix $\hat{Y}$. Decompose the formula (4), (5) into direct and indirect parts, and get:

$$\hat{X}_G = \hat{G} \times L \times \hat{Y} = \hat{G} \times I \times \hat{Y} + \hat{G} \times (L - I) \times \hat{Y} = \Delta X_G + \Delta \hat{X}_G \quad (6)$$

$$\widehat{X}_F = \widehat{F} \times L \times \widehat{Y} = \widehat{F} \times I \times \widehat{Y} + \widehat{F} \times (L - I) \times \widehat{Y} = \Delta X_F + \Delta \widehat{X}_F \tag{7}$$

Where

$$\widehat{E} = \begin{bmatrix} e_{11} \cdots e_{1n} \\ \vdots \quad \vdots \\ e_{nn} \cdots e_{nn} \end{bmatrix}, \widehat{F} = \begin{bmatrix} f_{11} \cdots f_{1n} \\ \vdots \quad \vdots \\ f_{n1} \cdots f_{nn} \end{bmatrix} \text{ and } \widehat{Y} = \begin{bmatrix} y_{11} \cdots y_{1n} \\ \vdots \quad \vdots \\ y_{n1} \cdots y_{nn} \end{bmatrix}$$

In the formulas, ΔX_G and ΔX_F refer to the direct increase in value and direct energy consumption value resulting from investment respectively. $\Delta \widehat{X}_G$ and $\Delta \widehat{X}_F$ refer to indirect increment in value and indirect energy consumption resulting from investment respectively.

2.2 Data sources

2.2.1 Input-output table

We construct a non-competitive input-output table 2015 according to the methods of Tian et al. (2018) , which could eliminate the impact of imported goods and services (Su and Ang, 2013). The total capital formation in China is the final demand Y for investment in the construction industry after excluding imported products. In addition, according to the added value in the 2015 input-output extension table and the "China's Energy Statistical Yearbook 2015", the economic added value and energy consumption intensity of each sector were obtained.

2.2.2 Sector classification

In order to match the classification of each sector with the energy consumption data in China Energy Statistics Yearbook 2016 (Chen and Zhang, 2010), we divided all sectors into the following 35 sectors.

Table 1. Classification

No.	Sector	No.	Sector
sector1	Agriculture, forestry, animal husbandry and fishery	sector19	Electrical, machinery and equipment manufacturing
sector2	Coal mining and washing	sector20	Communication equipment, computer and other electronic equipment manufacturing
sector3	Oil and gas extraction	sector21	Instrumentation and cultural office machinery manufacturing
sector4	Metal Ores Mining and Dressing Industry	sector22	Other manufacturing
sector5	Nonmetallic Mineral Mining and Dressing Industry	sector23	Scrap
sector6	Mining support services and other mining products	sector24	Repair of metal products, machinery and equipment
sector7	Food manufacturing and tobacco processing	sector25	Production and supply of electricity and heat
sector8	Textile industry	sector26	Gas production and supply
sector9	Clothing leather down and its products	sector27	Water production and supply
sector10	Wood processing and furniture manufacturing	sector28	Construction industry

sector11	Paper printing and stationery manufacturing	sector29	Transportation and warehousing
sector12	Petroleum processing, coking and nuclear fuel processing industries	sector30	Postal industry
sector13	Chemical industry	sector31	Wholesale and retail trade
sector14	Non-metallic mineral products industry	sector32	Accommodation and Catering
sector15	Metal smelting and rolling processing industry	sector33	Finance and insurance
sector16	Metal products industry	sector34	Real estate
sector17	General and special equipment manufacturing	sector35	Integrated Services
sector18	Transportation equipment manufacturing		

3 Results

3.1 The effect of construction investment on GDP and energy consumption

The development of the construction industry consumes lots of materials and gravel, energy, labor, etc. Simultaneously, the construction industry, transportation equipment manufacturing, special and general equipment industries, various public facilities management industries, and the provision of diversified services various departments of the national economy, such as scientific research and technical service industries, have invested a lot of production and services for project construction, thereby stimulating the economic development of these industries. According to Figure 1, it can be seen that the investment in the construction industry in 2015 boosted economic growth of 15.11 trillion yuan, of which the economic growth value directly driven was 4.11 trillion yuan, accounting for 27.22%, and the indirect growth was 11.00 trillion yuan, accounting for 72.78 %.

The construction of various buildings involves the consumption of raw materials such as electricity, heat, cement, and steel. The input departments of these materials are all high energy consumption departments, which will further cause a large amount of energy consumption. According to Figure 2, it is known that the 1701.10Mtce energy consumption caused by the investment in the construction industry in 2015, only 20.98Mtce came from the direct energy consumption of the construction sector, accounting for only 1.23%, and the other 1680Mtce energy consumption came from the indirect development of the construction industry. Energy consumption accounted for 98.77%. Therefore, future policy development was supposed to focus on indirect energy consumption in the construction. This is in line with Giesekam (2015) and others who believe that the development of the construction industry will generate a large amount of indirect carbon emissions.

 https://doi.org/10.1051/e3sconf/202015501008

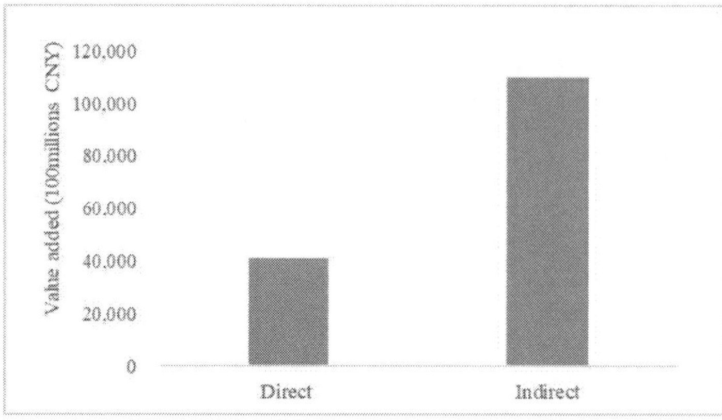

Fig. 1. GDP growth due to construction investment

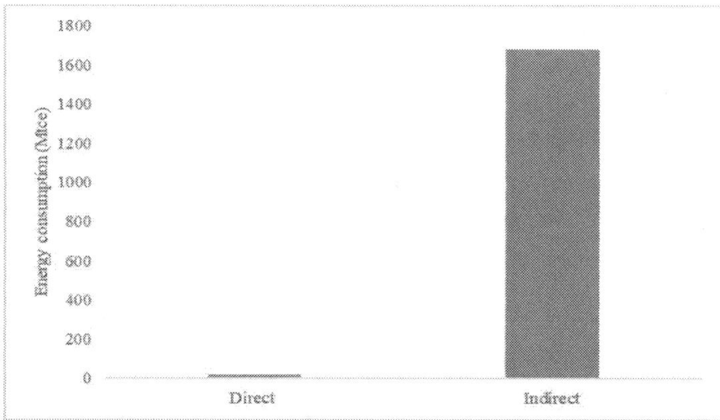

Fig. 2. Energy consumption growth due to construction investment.

3.2 The sectoral analysis of construction investment on GDP and energy consumption

According to Figure 3, the investment in the construction industry in 2015 can significantly boost the growth of the Construction, Integrated services, Financial Intermediation, Nonmetallic Mineral Products, and Wholesale and Retail Trades sectors. The economic growth value driven by these five sectors accounts for nearly 60%. It shows that the construction investment has a positive role in promoting the integrated service industry and the financial industry. On the one hand, the construction industry is a labor-intensive industry that can directly accommodate a large amount of labor. At the same time, its upstream related metal product manufacturing, machinery and equipment manufacturing, building materials, and transportation departments also absorb a large number of employees. Therefore, The construction industry indirectly accommodates a large number of employed people and contributes to alleviating employment pressure in China. On the other hand, because the construction industry has the characteristics of large investment, large resource consumption and long product formation time, for the purpose of ensure the stable development of the construction industry, China's banks, funds, trusts, insurance and other financial institutions have provided diversified financial products for the construction industry, which effectively revitalizing resources and realizing effective allocation of funds.

According to Figure 4, the sectors which have the largest energy consumption are production and supply of electricity and heat, Metal smelting and rolling processing industry, non-metallic mineral products industry, which respectively reached 1014.62 Mtce, 247.98 Mtce, and 176.52 Mtce. The total proportion reached 84.6%, indicating that the industry's policy making efforts in these three sectors should be increased in the future. The largest proportion of the Production and Supply of Electricity and Steam sectors indicates that China still relies on traditional production methods and has a greater demand for energy such as electricity and heat. There is an urgent need to find cleaner and more efficient energy production methods. This is similar to previous research results. For instance, Jindao Chen (2017) found that investment in the construction industry will generate more significant indirect energy consumption. However, some scholars have put forward different views. For example, Chuai et al. (2015) believe that energy consumption due to investment is mainly affected by production and supply of electricity and heat, water production and supply, metal products industry. Hong et al.(2016) found that the chemical industry is the most important sector of energy consumption in the construction industry.

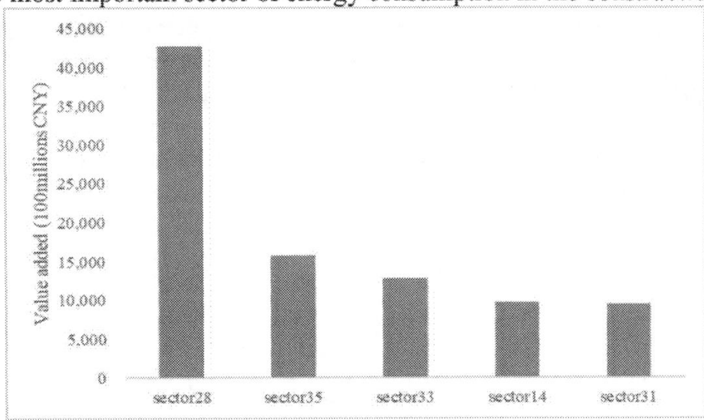

Fig. 3. The GDP growth of sectors due to construction investment

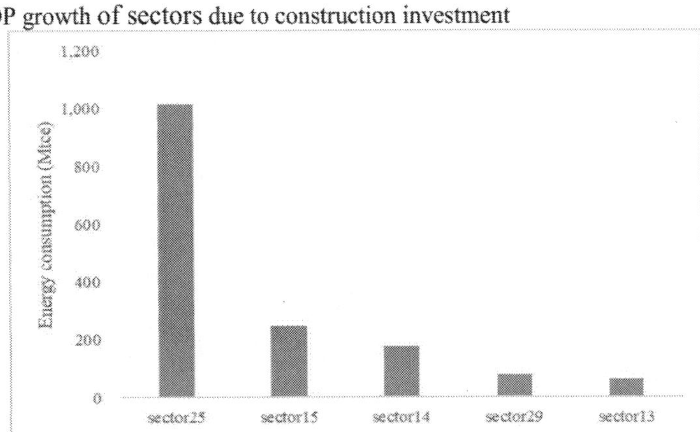

Fig. 4. The energy consumption of sectors due to construction investment

4 Conclusions

Our analysis focuses on the impact of China's construction investment in 2015 on China's economy and energy. By building the input-output model, it is known that the economic growth value of the construction industry investment in 2015 was 15.12 trillion yuan, of which the direct economic growth value was 4.11 trillion yuan, and the indirect economic growth value was 11.01 trillion, accounting for 27.22% and 72.78% respectively. The energy consumption caused was 1701.10Mtce, of which direct energy consumption was 20.98Mtce and indirect energy consumption was 1681.12Mtce, accounting for 1.23% and 98.77% respectively.

In addition, construction, integrated services, and financial intermediation are the top three sectors in all sectors that are most affected by investment in the construction industry, which account for 47.09%. The top three sectors that can generate the most energy consumption are Production and Supply of Electricity and Steam, Smelting and Pressing of Metals, Nonmetallic Mineral Products, which together account for up to 84.60%.

On this basis, we give the following policy recommendations:

First of all, the sectors which have the largest energy consumption due to investment in the construction industry are concentrated in traditional energy-intensive industries such as electricity and heat, indicating that China has a strong dependence on traditional energy. From the high-frequency use of non-metallic mineral products in the construction industry, we can also find that China uses more coal-consuming products. In the future, efforts should be made to optimize the energy consumption structure.

Secondly, according to the calculation results, it can be known that the indirect energy consumption due to construction investment is far greater than the direct. Therefore, future emission reduction policies were supposed to focus on the indirect energy consumption generated during the development of the construction industry. For example, improving traditional technologies, reducing material waste during construction, and increasing material usage during construction. These methods can help policymakers effectively cut back indirect energy consumption when investing in the construction industry.

References

[1] Chaudhri D P,Savings, investment, productivity and economic growth of Australia 1861-1990: some explorations. Economic Record, 76(232):55-73 (2000)

[2] Gozgor, G., Energy consumption and economic growth: New evidence from the OECD countries. Energy, 153: p. 27-34 (2018)

[3] Zhang, Q., Does one path fit all? An empirical study on the relationship between energy consumption and economic development for individual Chinese provinces. Energy,. 150: p. 527-543 (2018)

[4] Bakirtas, The relationship between energy consumption, urbanization, and economic growth in new emerging-market countries. Energy, 147: p. 110-121 (2018)

[5] Feng Qian, Hou Hexin, Li Dongyu, Foreign Direct Investment, Domestic Capital and Economic Growth in Economic Transition: Econometric Tests and Empirical Analysis Based on Chinese Time Series Data. Enterprise Economy, (4): 38-41 (2012)

[6] Hui Yan, Linda C.H.Fan,Building and environment,Qinping Chen. Greenhouse gas emissions in building construction: A case study of One Peking in Hong Kong[J] , (45)949-955 (2010)

[7] Malindu Sandanayake. Environmental emissions at foundation construction stage of buildings-Two case studies[J]. Building and environment, 95(01):189-198 (2016)

[8] Liu, L. Changes to pollutants and carbon emission multipliers in china 2007–2012: an input-output structural decomposition analysis. J ENVIRON MANAGE 203, 76-86. doi:10.1016/j.jenvman.07.061 (2017)

[9] Chen, J., et al., An empirical study on the CO 2 emissions in the Chinese construction industry. Journal of Cleaner Production, 168: p. 645-654 (2017)

[10] Tian, Y. Structural path decomposition of carbon emission: a study of China's manufacturing industry. J. Clean. Prod. 193, 563e574 (2018)

[11] Su, B. Input-output analysis of CO2 emissions embodied in trade: competitive versus non-competitive imports. Energy Policy 56, 83e87 (2013)

[12] Chen, G.Q. Greenhouse gas emissions in China 2007: inventory and input-output-analysis. Energy Policy 38, 6180e6193 (2010)

[13] Giesekam.Construction sector views on low carbon building materials. Building Research & Information, p. 1-23 (2015)

[14] Chuai, et al., Spatiotemporal Changes of Built-Up Land Expansion and Carbon Emissions Caused by the Chinese Construction Industry. Environmental science & technology, 49: p. 13021 (2015)
[15] Hong, J., et al., Energy use embodied in China's construction industry: A multi-regional input–output analysis. Renewable and Sustainable Energy Reviews, 53: p. 1303-1312 (2016)

Stochastic optimization of carbon mitigation path in Shenzhen based on uncertainty of power demand

Qinghui Hu[1], *Yushen* Tian [2,*]

[1,2]School of Environment and Energy, Peking University , Nanshan District, Shenzhen, China

Abstract. The core elements of urban carbon emission mitigation optimization path include structural adjustment, low energy supply, technological innovation, and enhanced energy demand management and improvement. How to optimize the combination of these factors to achieve the city's emission mitigation goals at the lowest cost is very important to study the path of urban low-carbon development. Due to many factors involved, it is difficult to solve this problem by building a mathematical optimization model that includes all the elements. This paper minimizes the total cost of emission mitigations in various departments in Shenzhen, combines the uncertainty of parameters and constraints, and uses mathematically standardized method to establish a stochastic optimization model for urban carbon emissions paths. Considering the uncertainty of energy demand, the optimal promotion rate of technical measures of the city's various departments in the stochastic optimization model during the planning period can be obtained, and the optimal solution of the city's low-carbon development optimization path can be formed.

1 Introduction

With the acceleration of urbanization and industrialization, China's energy demand and consumption have gradually increased, and the contradiction between energy supply and demand has become increasingly intensified. Among all industries in the city, the carbon dioxide emissions of the power, manufacturing, transportation, and construction sectors are relatively large. At the same time, these industries also have large emission mitigation potentials. To achieve the urban energy conservation and emission mitigation goals, the key is to reduce carbon emissions of these industries, but the promotion of emission mitigation technology requires capital investment. While promoting energy conservation and carbon emission mitigation, we need to control emission mitigation costs. Therefore, how to achieve the maximum emission mitigation at the lowest cost is worthy of further study.

Many scholars have studied the carbon emission mitigation path through model. The research of foreign scholars mainly focuses on the power sector. Zhang et al. [1] established a multi-period model and optimization framework for China's power industry. Jin et al. [2] developed Carbon Capture and Storage technology based on carbon mitigation targets and

* Corresponding author: tianyushen1211@pku.edu.cn

minimizing construction costs. Shi et al. [3] proposed energy systems integration considering the uncertainty of energy supply and demand. Cheng et al. [4] considered changing the investment decision to make a development plan for the production period from 2011 to 2050. Lee [5] presented a multi-period power plant planning based on carbon capture (CC) technologies. Zhang et al. [6] proposed a multi-period optimization model of China's power sector by considering the uncertainties of power demand, fuel prices and carbon tax policy. Some domestic scholars also research the development plan of the power sector. For example, Qiu et al. [7] presented a grid planning model that satisfies the system reliability criteria as the constraint and the construction costs of each stage of the system development process as the optimization goal. At the same time, some scholars turned their research objects to other departments. Ma et al. [8] used the optimization mathematical method and explored the optimal path of carbon emission mitigation in Shenzhen manufacturing. Tang et al. [9] selected 42 sectors to build a input-output model of multi-objective based on maximizing economic development and minimizing carbon emissions as optimization goals.

However, due to the availability of data, there is no article that studies carbon emission mitigation technologies in detail, and related research often focuses on one sector, and lacks overall research on multiple sectors. At present, Shenzhen is forming a win-win situation for environmental protection and economic development. Studying Shenzhen's low-carbon development path lends reference to other cities. Therefore, this paper establishes a stochastic optimization model with the goal of minimizing the total cost of emission mitigation in four sectors of Shenzhen by using the promotion rate of each carbon emission mitigation measure as an optimization variable. Combining the uncertainty of parameters and constraints, mathematical methods are used to determine the theoretical optimal values. After the theoretical optimal values of the promotion rate of emission mitigation measures in four sectors are calculated, the theoretical optimal solution of the emission mitigation path is determined, which can be used as a theoretical basis for urban carbon emission mitigation work and set energy saving and emission mitigation targets.

2 Materials and Methods

2.1 Model Building

The structure of the optimization model mainly includes three major elements: objective function, decision variables, and constraints. The objective function of the urban carbon emission mitigation optimization model is to minimize the total cost of each department during the planning period, and the decision variable is the promotion rate of each department's technology and other measures. Constraints mainly include carbon emission constraints, fuel supply constraints and non-negative constraints.

For the convenience of description, the physical meaning of each variable in the model is given in Table 1 in the appendix.

The objective function of the urban carbon emission mitigation path optimization model is to minimize the cumulative costs of the power sector, manufacturing sector, transportation sector, and construction sector during the planning period, as shown in equation (1):

$$C=\sum_{t=2015}^{2030}\left[\frac{\sum_{1}^{n_1}(tinv_i^{nb}+tinv_i^{rf}+tom_i+tfu_i+tex_i)*x_i+\sum_{n_1+1}^{n_2}tinv_i^{rf}*x_i+\sum_{n_2+1}^{n_3}tinv_i^{rf}*x_i+\sum_{n_3+1}^{n_4}tinv_i^{rf}*x_i}{(1+I)^{t-2015}}\right] \quad (1)$$

The constraint conditions are shown in the following formula:

(1) Carbon mitigation constraints. Equations (2) to (5) indicate that the total carbon emission mitigation of each technology in different sectors should be greater than or equal to the planned carbon emissions:

$$\sum_{t=2015}^{2030} \sum_{i=1}^{n_1} \sum_{f} x_i * ic_i * sav_{i,f} * ef_f \geq F_{t,e}^{pl} \quad (2)$$

$$\sum_{t=2015}^{2030} \sum_{i=n_1+1}^{n_2} x_i * sav_i \geq F_{t,m}^{pl} \quad (3)$$

$$\sum_{t=2015}^{2030} \sum_{i=n_2+1}^{n_3} x_i * sav_i \geq F_{t,c}^{pl} \quad (4)$$

$$\sum_{t=2015}^{2030} \sum_{i=n_3+1}^{n_4} x_i * sav_i \geq F_{t,t}^{pl} \quad (5)$$

(2) Electricity demand constraints. Due to uncertainty of many factors such as the speed of economic growth and future power demand, ε_{ED} is set to be the stochastic fluctuation of electricity demand, then we can express the power demand constraint more reasonably:

$$\sum_{k=1}^{m_1} ic_{k,t} * OH_k + \sum_{k=m_1+1}^{m_2} x_k * IC_{k,t} * OH_k + Su_{ex} \geq ED_t + \varepsilon_{ED} \quad (6)$$

For the stochastic fluctuation ε_{ED}, we can use the opportunity constrained programming model to solve it, this model assumes the probability that the constraint is true is greater than or equal to a certain confidence level:

$$\Pr\left(\sum_{k=1}^{m_1} ic_{k,t} * OH_k + \sum_{k=m_1+1}^{m_2} x_k * IC_{k,t} * OH_k + Su_{ex} \geq PD_t + \varepsilon_{PD}\right) \geq \alpha \quad (7)$$

(3) Installed capacity constraints. Equation (8) indicates that each type of power plant's installed capacity should not exceed its own maximum $IC_K^{\max}$:

$$x_k * IC_K \leq IC_K^{\max} \quad (8)$$

(4) Fuel supply restrictions. Equation (9) states that each fuel's supply should not exceed its maximum $FSC_{f,t}^{\max}$:

$$\sum_{k=1}^{n} x_k * fd_{f,t}^k \leq FSC_{f,t}^{\max} \quad (9)$$

(5) Non-negative constraints. As shown in Equation (10), the promotion rate of various carbon emission mitigation technologies in four sectors should be set between 0 and 1.

$$0 \leq x_i \leq 100\%, i = 1, 2, \ldots \quad (10)$$

2.2 Empirical analysis——taking Shenzhen as an example

2.2.1 Carbon mitigation measures

According to the specific situation of Shenzhen, Carbon mitigation technologies in different sectors are shown in Table 2 in the appendix.

2.2.2 Data Sources

First, the list of emission mitigation technologies and indicator for four sectors was collected through field research, enterprise questionnaire survey, expert consultation and literature research. Secondly, other parameters of the power sector were obtained through field surveys from different power plants in Shenzhen. Finally, the planned emission mitigation was calculated according to the outline of the 13th five-year plan for Shenzhen's national economic and social development, the original carbon emission data of Shenzhen and the forecast of Shenzhen's economic growth.

3 Results and Discussion

According to the model and data proposed in Section 2, the optimal carbon emission mitigation paths for Shenzhen's power sector, manufacture sector, construction sector and transportation sector under uncertain power demand are obtained.

3.1 Power Sector

Figure 1 shows the promotion rate of various emission mitigation technologies in Shenzhen's power sector from 2015 to 2030.

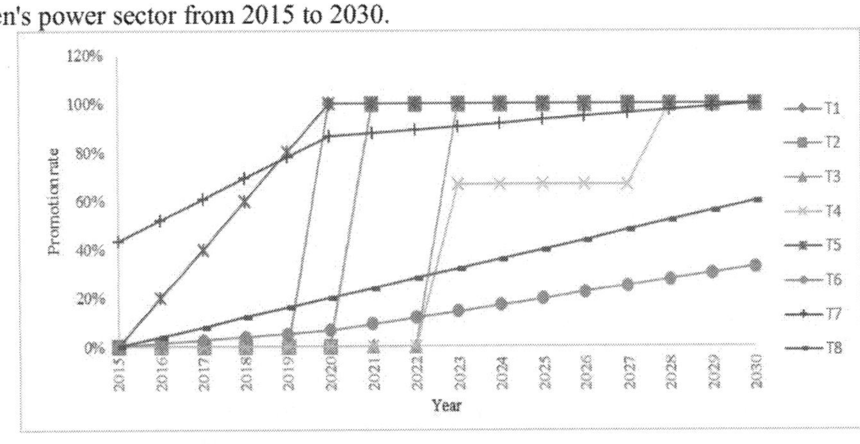

Fig. 1. Promotion rates of different technologies in the power sector

We found that some technologies have the same popularization rate, so different technologies can be combined into 8 subsets, as shown in Table 3. in the appendix.

As shown in Figure 3.1, The promotion rate of technology in subset T1 will first reach 100% in 2020, subsets T2 and T3 are implemented immediately after T1. This is because the carbon emission mitigation amount and cost of different technologies are different. The cost of T1 is relatively low so the carbon mitigation is more efficient. Distributed photovoltaic power generation(T6) is a distributed power generation system that can directly convert solar energy into electrical energy. It is a new type of power generation and comprehensive energy utilization method with broad development prospects. It is limited by the high cost and developed relatively slowly. Waste-to-energy(T7) is a good way to utilize waste. Its development is relatively fast because of benefiting from the government subsidies, which has also accelerated the decommission of the Mawan Power Plant.

3.2 Manufacturing, Construction and Transportation sectors

The promotion rates of various emission mitigation technologies in the manufacturing, construction and transportation sectors from 2015 to 2030 are shown in Figure 2, Figure 3 and Figure 4.

Manufacturing technology is integrated into three subsets, subset T9 includes i31, T10 includes i46, and T11 includes the remaining technologies for manufacturing carbon mitigation. Different technologies in T11 have the same promotion rate, mainly because their carbon emission mitigation efficiency is relatively low. As shown in Figure 3.2, the technology in subset T10 will be implemented in 2028. Taking into account the discount rate, the amount of investment will decrease over time, which makes the carbon emission mitigation efficiency of T10 continue to improve to be close to T9 in 2028. The technology in T11 does not improve carbon emission mitigation efficiency over time, so the promotion rate has always been zero.

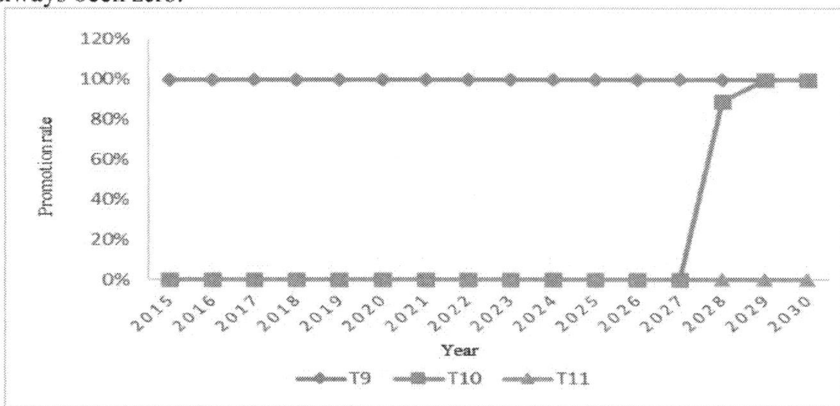

Figure 2. Promotion rates of different technologies in the Manufacturing sector

The technology of the construction sector is integrated into 3 subsets, the subset T12 includes i80, i86, i105, i106, i107, T13 includes i77, and T14 includes the remaining carbon emission mitigation technologies of the construction sector. The promotion rate of T12 technologies during the planning period is 100%, because their investment is relatively low. For example, the cost of the Energy Monitoring Statistics Database is close to zero, so its carbon emission mitigation efficiency is also higher. As shown in Figure 3.3, the technology in the subset T13 will be improved quickly in 2030, and the carbon emission mitigation efficiency of T13 in 2030 will gradually approach that of T12. The technology in T14 does not improve carbon emission mitigation efficiency during the planning period, so the promotion rate has always been zero.

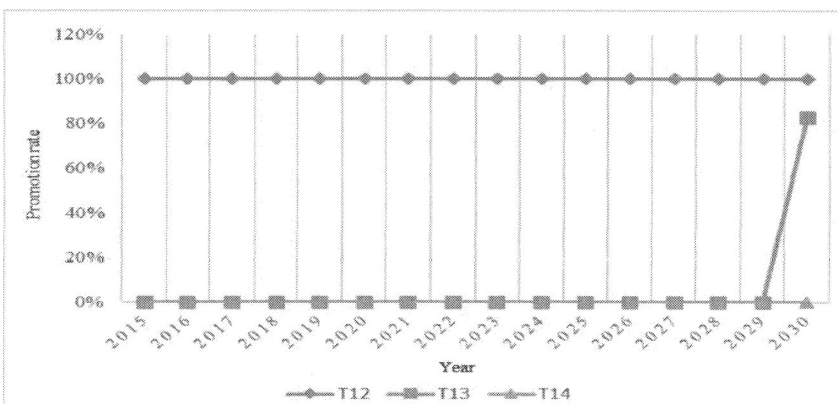

Figure 3. Promotion rates of different technologies in the Construction sector

The technology of the transportation sector is integrated into two subsets. Subset T15 includes i120, i124, i129, i130, i131 and i134-i141, and T16 includes the remaining emission mitigation technologies of the transportation sector. The promotion rate of T15 technologies during the planning period is 100%, because their cost is zero, which also makes carbon emission mitigation efficiency very high. Reducing carbon emissions by managing energy conservation and other zero-cost methods will be an important means of reducing carbon emissions in the transport sector.

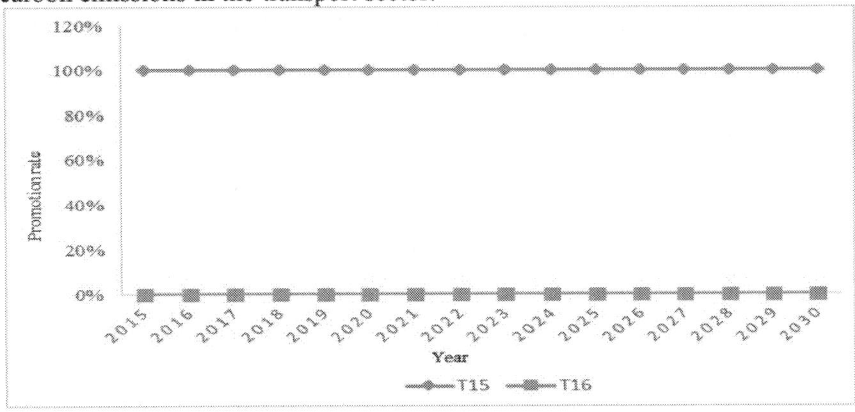

Figure 4. Promotion rates of different technologies in the Transportation sector

4 Conclusions

In this paper, considering the uncertainty of power demand and the background of the low-carbon development plan of Shenzhen, a mathematical planning method is used to establish the stochastic optimization model of urban carbon emission paths. We calculate the promotion rate of each emission mitigation measure during the planning period with the goal of minimizing the total cost of reducing emissions in four sectors. Then it can determine the theoretically optimal solution for each emission mitigation subdivision path, which will lay a foundation for the design and evaluation of low-carbon development routes. It also points out the direction for the actual setting of energy-saving emission mitigation targets and the implementation of energy-saving emission mitigation measures.

The empirical analysis of Shenzhen shows that the best promotion strategy for urban emission mitigation measures during the planning period is to promote the emission mitigation measures with the lowest emission mitigation cost and the highest emission mitigation efficiency, and to make their promotion rate reach 100% as soon as possible. For example, many of the measures preferentially promoted in the transport sector are low-cost policy technologies, on the contrary, the promotion rate of emission mitigation measures with high emission mitigation costs is low, such as T16 emission mitigation measures in the transportation sector. According to this stochastic optimization model, we can get the optimal promotion rate of the technical measures of four sectors in the city during the planning period, and form the optimal solution of the low-carbon development path under certain exogenous conditions. The stochastic model guarantees the robustness of the results and the stability of policy recommendations.

If we promote various energy saving and emission mitigation technologies in accordance with the promotion rate calculated from the optimization model, the planned emission mitigation goals will be completed at the lowest cost during the planning period. However, since the promotion of emission mitigation measures is affected by policies, promotion costs and other factors, the promotion of various measures is often different from the ideal situation. In the actual promotion, the city's specific conditions and the efficiency of energy-saving and emission mitigation measures should be fine-tuned to design an optimal path of carbon emission mitigation in line with the city's actual conditions.

Acknowledgements

This work did not receive any specific grant from funding agencies in the public, commercial, or not-for-profit sectors.

References

1. D. Zhang, P. Liu, L. Ma, Z. Li, W. Ni, COMPUT CHEM ENG,**37**(2012)
2. Jin, L. Bai, J. Kim, S. Jeong, K. Kim, ENERGIES,**10**(2017)
3. J. Shi, Y. Wang, R. Fu, J. Zhang, SUSTAINABILITY-BASEL,**9**(2017)
4. R. Cheng, Z. Xu, P. Liu, Z. Wang, Z. Li, I. Jones, APPL ENERG,**137**(2015)
5. J.Y. Lee, APPL ENERG,**198**(2017)
6. D. Zhang, P. Liu, L. Ma, Z. Li, COMPUT CHEM ENG,**50**(2013)
7. **W. Qiu,** Y. Qiu, Zhejiang Electric Power,29(2010)
8. **X. Ma,** B. Hu, H. S, J. J, Modern Management Science, **11**(2016)
9. J. T, B. Z, S. C, Journal of Industrial Technological Economics,**12**(2012)

 https://doi.org/10.1051/e3sconf/202015501010

Liquefaction performances of bread crusts in subcritical water

Yuzhen Wang*, *Zhuan* Liu, *Changqing* Fang, *Zeyu* Zhang, *Li* Xie

Shaanxi Provincial Key Laboratory of Printing and Packaging Engineering, Xi'an University of Technology, Xi'an, Shaanxi, 710048, China

Abstract. Hydrothermal liquefaction of bread crusts in subcritical water were performed in a micro-batch reactor. The influences of temperature (300 – 360 °C) and residence time (10-30 min) on bio-oil yield, boiling point distribution and functional groups in bio-oil were investigated. The results showed that bio-oil yield increased with increasing temperature and reaction time. Maximum bio-oil yield of 22.69wt% was obtained at 360 °C, 30min. The longer reaction time promoted the degradation of diesel to jet fuel and naphtha. The naphtha (C7-10) and jet fuel (C11-C15) increased to 29.9%, 51.82% at 30min from 20.49% and 36.14%, respectively. FT-IR analysis showed that esters, ketones, amides, acids and aldehydes were present in the bio-oil.

1 Introduction

Food waste is solid or liquid organic waste produced in the process of daily diet. Fermentation and decay will occur at the action of microorganisms. If the food waste is not properly treated, it will not only harm people's lives and health, but also cause environmental pollution and pollute the atmosphere and water sources. The main treatment process is landfill, however, it covers large amount of land, and is prone to create secondary pollution, such as the landfill leachate[1].Hydrothermal liquefaction (HTL) in subcritical water is a process that can directly convert wet organic wastes into bio-oil avoiding the energy-intensive and expensive dry step required in pyrolysis[2] . As the main component of food waste is the cereals, which are mainly composed of starch, we selected bread crusts as the research object, the effects of the reaction temperature and time on the bio-oil and solid residue yield were studied. And also the components of the bio-oil were investigated.

2 Experimental

2.1 Materials

* Corresponding author: yzwang@xaut.edu.cn

The bread crusts was first dried in the oven at 80 °C, and then was grinded to about 20-30 μm.

2.2 Procedures and separation of the products

The experiments were performed in a micro-reactor. The reactor is made of 316L stainless steel, and is designed to withstand a maximum temperature and pressure of 480 °C and 38MPa, respectively. A certain amount of bread crusts and deionized water were charged into the reactor, and then the sealed reactor was placed in a tube furnace to heating to the reaction temperature. After the reaction time was reached, the reactor was taken out and cooled in cold water immediately. The experimental process was shown in Fig.1. The products in the reactor were washed with 9 ml dichloromethane, and then the products was subjected to solid-liquid separation by a centrifuge. Finally, the dichloromethane in the solution was blown away with a nitrogen blower, and the rest liquid was bio-oil. The separated solid residue was dried and weighed in an oven at 80 °C.

The bio-oil yield and residue yield are calculated as the mass ratio of bio-oil to reacted bread crusts, and ratio of solid residue bio-oil to reacted bread crusts, respectively.

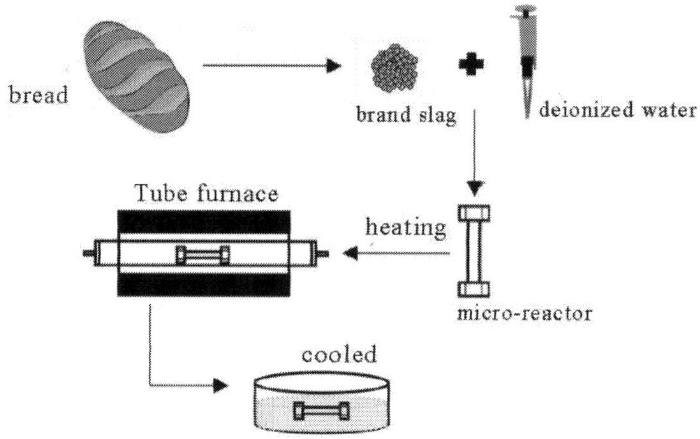

Fig.1 Experimental procedures

2.3 Analysis

Thermogravimetry analysis (TGA) (209 F3 Tarsus, Nietzsche, Germany) was used to evaluate the boiling point distribution of the bio-oil. Samples were taken at a temperature of 25°C to 550°C in an N_2 atmosphere of 60.0ml/min at 10°C/min. Fourier infrared spectrometer (FT-IR) (Shimadzu FTIR-8400S) was used to determine the functional groups contained in the bio-oil, the measurement frequency range is 4000-500cm^{-1}.

3 Results and discussions

3.1 Effect of temperature and reaction time on bio-oil yield

The effects of different reaction temperature and reaction time operated at 20MPa on the bio-oil yield and residue yield of bread residue in subcritical water are shown in Fig.2. The reactant concentration kept at 20wt%. It can be clearly observed that the bio-oil yield increased obviously with the increasing temperature from 300°C to 360°C. And the solid residue decreased accordingly. The bio-oil yield at 300°C, 330°C and 360°C were 15.77%, 20.39%, and 22.69%, at the reaction time of 30min, respectively. Higher temperature promoted the degradation of starch and glucose in the bread crusts [3,4,5].

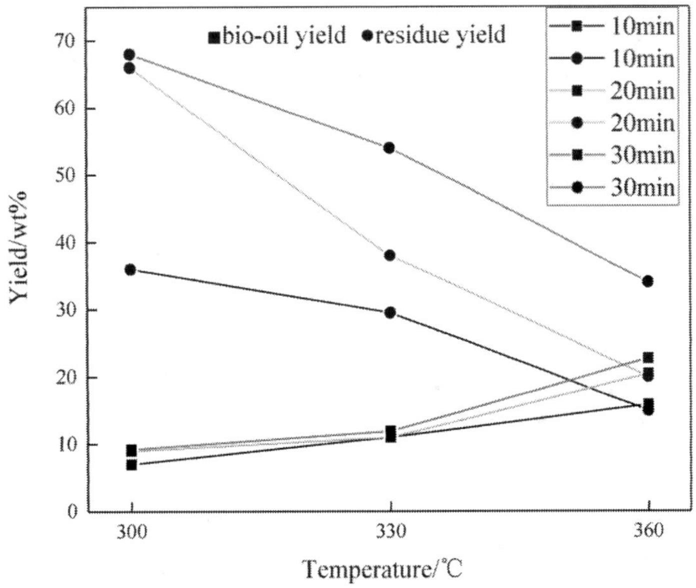

Fig.2 Changes of bio-oil and residue yield with temperature and reaction time

In the reaction time range of 10-30min, the bio-oil yield increased with the reaction, this means that the reaction was not completed before 30min. The variation trend of residue yield was opposite. Similar results were also found in the HTL of swine manure [4], military surrogate waste [1] and Dunaliella salina [6].

3.2 Boiling point distribution in the bio-oil

The boiling point distributions of bio-oil were evaluated by the TGA, which can be viewed as a miniature "distillation" [7]. Fig.3 showed the boiling point distribution and thermal conversion properties of bio-oil operated at 360°C for 20 min and 30 min. The fuel fractions with the boiling point range of 100 -200 °C, 200-300 °C, and 300-550 °C, >550 °C were defined as naphtha, jet fuel, diesel and vacuum residue, respectively, as shown in Table 1. Results showed that the naphtha(C7-10) and jet fuel (C11-C15) increased greatly at 30 min compared with that at 20min, and the value climbed from 20.49% to 29.9%, and 36.14% to 51.82%. While the content of diesel (C16-C47) decreased from 41.78% to 16.97% from 20min to 30min. The results indicated that large molecular compounds were degraded to smaller molecular compounds with the increasing reaction time. The content of vacuum residue (>C48) were very low at 20 min and 30 min, which mean that the intermediates rarely polymerized to the coking products.

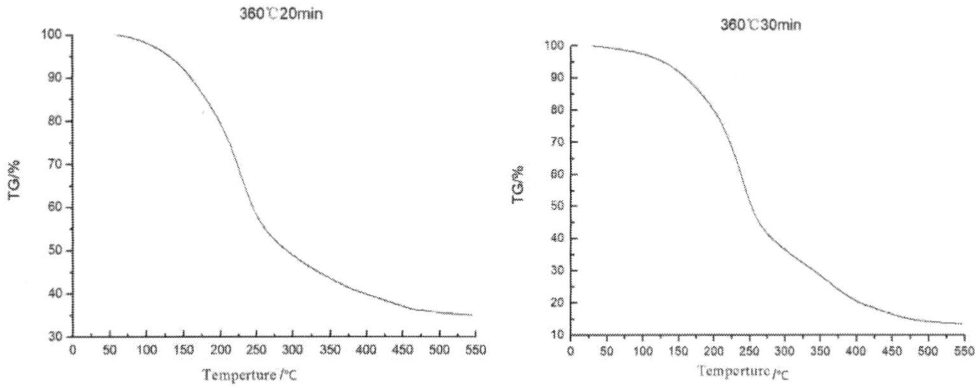

Fig.3 TG curves of bio-oil

Table 1 Boiling point distribution of bio-oil

Distillate fraction / Time/min	Naphtha (<200 °C) C7 – C10	Jet fuel (200-350 °C) C11 – C15	Diesel (350-500 °C) C16– C47	Vacuum residue (>500 °C) >C48
20	20.49%	36.14%	42.78%	0.47%
30	29.90%	51.82%	16.97%	0.73%

3.3 Functional groups in the bio-oils

The FT-IR of bio-oil is shown in Fig.4. It can be seen that the temperature and reaction time showed little influence on the vibration peaks of the infrared spectrum, which indicating that the types of functional groups in the bio-oil are similar. The vibration peaks at 3200-3500cm^{-1} indicating the O-H groups, which reflected the presence of amines, phenols, and alcohols. The 2800-2950cm^{-1} represents the stretching vibration peak of C-H. Around 1700cm^{-1} is the C=O stretching vibration peak, which means that esters, ketones, amides, acids and aldehydes may present. Around 1600cm^{-1} and 1520cm^{-1} should be the bending vibration peaks of N-H and -NH_2, which represent amides and amines. Around 1500cm^{-1} is the C=C stretching vibration peak, which represents the benzene rings. Around 1380cm^{-1} and 1460cm^{-1} are -CH_2 bending vibration peaks and -CH_3 bending vibration peaks, which represent saturated C-H. 1300cm^{-1} to 1000cm^{-1} are C-O stretching vibration peaks, which are the strongest peaks in this region. They are easier to identify and represent phenols, acids, esters, and alcohols.

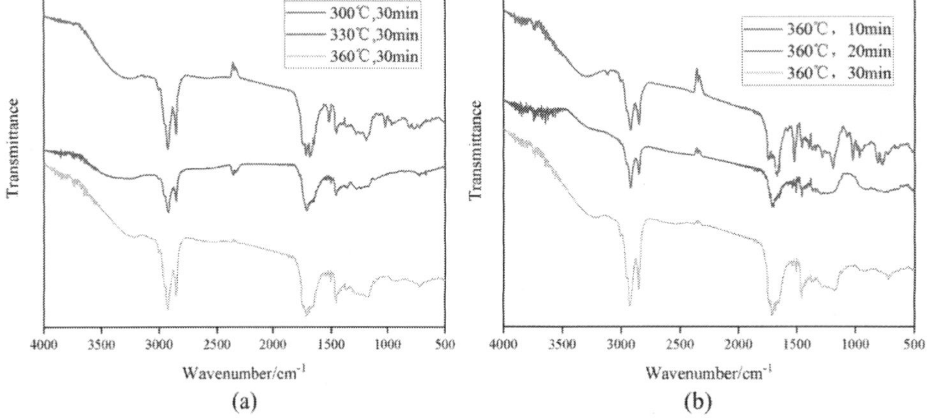

Fig.4 FT-IR spectra of bio-oil obtained at (a) 300-360°C, 30min (b) 360°C, 10-30min.

4 Conclusions

The hydrothermal liquefaction of bread crusts were operated in the micro-reactor at 300-360 °C and 10-30 min. The effects of temperature and reaction time on the bio-oil yield and components of the bio-oil were investigated. The results showed that temperature and reaction time showed positive effect on the bio-oil yield. At 360 °C, 30 min, maximum bio-oil yield of 22.69wt% can be obtained. Longer reaction time promoted the degradation of diesel to jet fuel and naphtha. The naphtha (C7-10) and jet fuel (C11-C15) increased to 29.9%, 51.82% at 30min from 20.49% and 36.14%, respectively. FT-IR analysis showed that esters, ketones, amides, acids and aldehydes were present in the bio-oil.

References

[1] D.J. Zastrow, P.A. Jennings, Florida Institute Techno (2013).
[2] R. Chand, V.B. Borugadda, M. Qiu, A.K. Dalai, Appl. Energy , 254. 113679(2019).
[3] Y. Qu, Z.D. Liu, Z.B. Zhu, et al. J. Solar. Energy. 37, 1333(2016).
[4] G.Yu, Y. Zhang, L. Schideman, et al. Transactions of the ASABE, 54, 239-246(2011).
[5] Y.H. Chan, A.T. Quitain, S.Yusup, et al. J. Energy. Inst, 91(5): 721-732(2018).
[6] B. Eboibi, D. Lewis, P. Ashman.Bioresour. Technol. 170, 20-29(2014).
[7] W.T. Chen, Y.H. Zhang, J.X. Zhang, G .Yu, L.C. Schideman, P.Zhang, et al. Bioresour Technol 152,130-9(2014).

Challenges and Policy Suggestions on the Development of Hydrogen Economy in China

Wei Shan[1,2], *Fang-Fang* Wang[3,*]

[1]School of Economics, Sichuan University, China

[2]School of Economics and Management, Wenzhou University Oujiang College, China

[3]School of Business, Wenzhou University, China

Abstract. In recent years, with the intensification of environmental damage, mankind is at a crossroads towards the next energy revolution. Hydrogen energy, because of its pollution-free nature, has become the focus of technological research and development in many countries, and the economic pattern around it is taking shape China's investment in the field of new energy has been increasing in recent years, and the prospect of hydrogen economy has prompted China to continuously increase the development of hydrogen energy. However, it faces many obstacles, both technical and social. The Chinese government is now using financial subsidies to help hydrogen companies incubate. However, with the development of the industry, China is currently facing great obstacles in large-scale commercial application, intellectual property protection and innovation, standardization construction and social confidence. This paper believes that policy measures such as selective commercial distribution, innovation system building, industry standardization construction and social education can be adopted to help solve these challenges

1 Introduction

Hydrogen energy, as a secondary energy, can be produced from a variety of primary energy, such as coal, oil, natural gas, solar energy, wind energy, hydropower, tidal energy, geothermal energy, nuclear fuel, etc. can be directly or indirectly produced hydrogen energy and then converted into electricity. Not so with secondary sources such as petrol and diesel, which are produced almost entirely on fossil fuels. With the increasing consumption of fossil fuels and the dwindling reserves of these resources, there is an urgent need to find a new and abundant energy source that does not depend on fossil fuels. Hydrogen energy is a kind of new secondary energy which is expected to solve the conventional energy crisis.

China currently has the world's largest hydrogen production capacity. China produced 16 million tons of hydrogen in 2012, rising to 20 to 22 million tons in 2018. Therefore, China has a good production base to meet the potential demand of the future hydrogen economy, known in China as the "hydrogen energy industry." According to the 2016 blue book on "Infrastructure Development of Hydrogen Industry in China", which compiled by

* Corresponding author: wff2005@126.com

China Institute of Standardization and Institute of Technology, by 2020, 72 billion cubic meters of hydrogen will be used for energy, and the industry's total industrial output will reach 300 billion yuan ($44.4 billion). This will increase further to 1 trillion yuan ($148 billion) by 2030 and 4 trillion yuan ($59.2 billion) by 2050.

Recently, the China Hydrogen Energy Alliance released its first white paper on hydrogen and fuel cell industry in China. The alliance founded in February 2018, and is led by China Energy Investment Corporation, and its members come from a number of universities and other well-known companies such as Linde, Air Products, Air Liquide. According to the document, the value of the hydrogen industry will reach rmb1tn ($148bn) between 2020 and 2025, with 50,000 fuel-cell vehicles in service and 200 refuelling stations. From 2026 to 2035, the industry's industrial output will jump to rmb5tn ($740bn) and 15m fuel-cell vehicles will be equipped with up to 1,500 refuelling stations. By 2050, hydrogen will account for 10 per cent of China's energy system, total demand will reach 60m tonnes and annual economic output will exceed rmb10tn ($1.48tn), the report added.

Despite the development of hydrogen economy is prosperous, the challenges are also evident. Based on the findings of Gong (2018) and the recently released statistical survey data, this paper makes a detailed analysis of the three challenges facing the development of hydrogen economy: (1) commercial scale application; (3) intellectual property protection and innovation; (4) safety standards and social confidence.

2 Challenges for Hydrogen Industry

2.1 Large-scale commercial application

At present, the large-scale commercial application of hydrogen energy still faces many challenges. In terms of hydrogen production, most of the existing hydrogen production technologies rely on primary energy sources such as coal and natural gas, and the economic and environmental problems are still prominent. Although the Hydrogen production and power generation system with zero net CO2 emission based on coal and supercritical water gasification proposed by Guo and Jin [1] effectively reduces carbon emissions, it is still considered to be pollution-prone to some extent [2]. The use of nuclear energy and biomass gasification to produce hydrogen is not yet mature, and the use of renewable energy such as solar energy or wind energy has low efficiency and high comprehensive cost [3-4].

In terms of hydrogen storage, although pressurized compression hydrogen storage technology, liquefied hydrogen storage technology, metal hydride hydrogen storage technology and organic compound hydrogen storage technology have all made great progress, the balance between hydrogen storage density, hydrogen storage safety and hydrogen storage cost has not been resolved, and there is still a certain gap from large-scale commercial application [5]. From the perspective of hydrogen use, the insufficient scale of hydrogen fuel cell vehicles leads to the high construction cost of hydrogen filling stations and difficulty in large-scale laying. In turn, the insufficient number of hydrogen filling stations makes it difficult for users to choose hydrogen fuel cell vehicles [6]. In general, it is difficult to combine the convenience and cost control of hydrogen industry.

2.2 Intellectual property protection and innovation

Developing hydrogen technology and infrastructure requires substantial investment, particularly in intellectual property, which can be an advantage for countries with strong innovation cultures and research capabilities. The trend is likely to shift countries' strategic focus from simply chasing energy supplies to strengthening infrastructure capacity and management system efficiency at the same time. As a result, in a world where renewable energy is the main source of energy, investment and technology have increasingly become sources of cooperation or nodes of geopolitical competition.

In the longer term, tensions over technology transfer between developing and developed countries have increased. Spending on renewable energy research among countries is a mix of geopolitical considerations as well as economic or environmental ones. Trade barriers disputes over photovoltaic cooperation between China and the EU or trade wars between China and the us are based on competition over intellectual property rights in energy technology are existing cases. In the future, countries will increasingly compete for renewable energy infrastructure. Asia has considerable renewable energy and accounts for a large share of the development of renewable energy. Whoever can seize the opportunity of renewable energy infrastructure in Asia will greatly enhance its geopolitical influence.

From the perspective of global hydrogen patent layout, a large number of core patents are in the hands of large enterprises in the United States, Japan and other countries, while China has not yet become the leading technology source of international hydrogen development. Most of the foreign patent applicants are multinational enterprises with strong strength, which have absolute leading technology advantages and capital advantages in the industry, while most of the domestic patent applicants are universities and scientific research institutes, which have weak applied technology basic research ability and low product transformation rate. In addition, foreign multinational enterprises, led by Toyota and Honda, have a strong awareness of patent protection and a clear strategic intention of international patent layout, and have applied for a considerable number of PCT patents in many countries. In contrast, Chinese research and development institutions lack awareness of patent protection and apply for fewer patents abroad, which is not conducive to future market competition and expansion.

2.3 Safety standards and social confidence

At present, the global academia and industry have reached a consensus that as long as the development of the hydrogen industry in accordance with the standards, the degree of safety can be guaranteed. According to incomplete statistics, taking water and hydrogen energy as an example, its security application as a new energy has been very mature. Compared with gasoline and diesel, hydrogen-water fuel has low calorific value, high flash point, not easy to ignite, high explosion limit, far less harmful than gasoline and diesel, high boiling point, high latent heat of vaporization, low volatilization, and lower on-site concentration than gasoline and diesel. Therefore, some scholars believe that filling stations with hydrogen fuel is safe, and vehicles can also ensure safety in driving. However, according to data from 100 consumers in Jiangsu province, the willingness to buy hydrogen cars in the near term is clearly low. The skeptical crowd, however, expressed widespread concerns about safety and a low level of trust in Chinese-made hydrogen products. But this concern is not groundless, at present China hydrogen safety technology investment is not much. It can be mainly reflected in the lack of high pressure hydrogen cycle fast charging and discharging fatigue test equipment, the lack of high sensitivity hydrogen leakage detection system and other key detection equipment, at the same time, more serious is the hydrogen safety detection research ability temporarily unable to meet the needs of hydrogen energy development. In addition to the shortcomings of the above standardization construction, this paper also believes that there are still large misunderstandings in the society's cognition of hydrogen energy. Alanne [7] once conducted a questionnaire survey on 93 Finnish students majoring in energy engineering to understand the cognition and cognition of hydrogen technology of college students. The results showed that the questionnaire before and after the homework improved students' knowledge and cognition of hydrogen technology. The biggest changes occurred in their knowledge of hydrogen safety and their willingness to acquire a home hydrogen system. However, the current Chinese society's coverage of various technologies is still not objective and enough. For example, after an explosion in Zhangjiakou in late 2018 that killed 23 people, media once linked the explosion to the hydrogen industry. Although later proved to be a rumor, but also caused public concern about the safety of hydrogen energy. Although China still needs to work hard on the establishment of hydrogen energy standardization, there are more than 21 hydrogen energy standards for the existing infrastructure in China, which are considered mature for the management of existing safety issues. So, how to build up society's confidence in hydrogen energy is still an urgent problem to be solved..

3 Policy suggestions

3.1 Selective Commercial distribution

The utilization of hydrogen energy still faces many obstacles, and too much political will and financial support will be detrimental to the development of hydrogen energy industry before major technological breakthroughs and significant cost reductions are achieved. The social and economic development status of our country cannot bear the huge cost of the hydrogen energy economy transformation which has not a clear prospect for a period of time. In view of the current situation in China, this study believes that the location conditions should be taken into consideration in the layout of hydrogen energy industry, and the development ideas and positioning should be clarified. Priority should be given to the development of hydrogen energy industry in such aspects as military use, space development, energy supply in remote areas, combination with renewable energy and small-scale commercial co-generation of heat and power.

Hydrogen can come from a wide range of sources, including hydrogen production from fossil fuels, hydrogen production from hydrogen-containing materials, hydrogen recovery from chemical byproducts, and hydrogen production from solar and wind energy. Hydrogen production from fossil fuels includes coal gasification to produce hydrogen, natural gas reforming to produce hydrogen and methanol cracking to produce hydrogen; Hydrogen production from hydrogen-containing substances includes hydroelectrolysis to produce hydrogen, ammonia decomposition to produce hydrogen and sodium borohydride hydrolysis to produce hydrogen; hydrogen recovery from chemical byproducts includes caustic soda, coke and light oil cracking.

Tian [8] believes that there are various production modes of hydrogen, and the economics of technology routes need to be analyzed according to regional resource endowment. Hydroelectrolysis is the process of injecting purified water into an electrolyzer with direct current. The water molecules undergo electrochemical reaction on the electrode and decompose into hydrogen and oxygen. Most of the cost of hydroelectrolytic hydrogen production is in electric energy. If there is cheaper off-peak electricity, the cost of hydrogen production will be greatly reduced. At present, Zhangjiakou, Hebei and other places have built a million kilowatt wind power demonstration base, due to wind abandon and power limit pressure, in Guyuan county, China's first comprehensive utilization of wind power hydrogen demonstration project. But for regions such as Datong in Shanxi Province, where coal reserves are abundant and coal chemicals are developed, hydrogen can be produced at a lower cost by gasification. In the long run, of course, technology should still be used to reduce the potential pollution from coal to hydrogen.
Adaptive Subsidy

At present, the hydrogen industry in China is very hot, but most regions lack the fuel cell vehicle industry, and there is also a lack of understanding of fuel cell vehicles. A considerable number of regions have introduced hydrogen industry support policies, but few have introduced specific policies for fuel cell vehicles, indicating that there is still a considerable distance from the industrialization of fuel cell. Of the eight regions that have introduced local subsidy policies, four have received land subsidy at a 1:1 ratio with the state subsidy, while only one region has received land subsidy at a ratio of less than 1:0.5. It is far higher than in 2018, all regions have attached great importance to fuel cell vehicles. This partly reflects the state's high subsidies for hydrogen, which have led to a rush of capital and a scramble to develop the industry in places that do not have the natural endowments of the "hydrogen economy". In this regard, this study encourages the development of market-oriented hydrogen energy industry, reasonably formulate subsidy policies, actively cultivate the market base, and promote the efficient development of hydrogen production, hydrogen storage and hydrogen use.

The localization rate of China's hydrogen energy industry chain is very fast. However, the primary problem of hydrogen energy industry is the cost, which needs to be reduced by mass quantification as soon as possible. However, the Ministry of Finance and other four departments jointly issued the notice on 'Further improving the financial subsidy policy for the promotion and application of new energy vehicles'. The transition period is from March 26, 2019, to June 25, 2019, according to the notice. After the end of the transition period, purchase subsidies for new energy vehicles (except buses and fuel cell vehicles) will no longer be provided, and the purchase subsidies will be concentrated on supporting the construction of weak links in infrastructure such as charging (hydrogenation) and supporting operation services. Long-term implementation of subsidy policy, so that some enterprises suffer from policy dependence, it is difficult for them to deal with global market competition. The latest subsidies remove some support for hydrogen development, and provides a strong signal for the future of hydrogen industry. The hydrogen industry urgently needs to provide more information to strengthen the subsidy determination of policy makers, at the same time, policy makers need to weigh the pros and cons and strengthen the strategic determination of subsidies. This paper suggests that, in the development of the hydrogen, the cancellation of the subsidy in a certain extent can stimulate the potential of the market. However, in the case that hydrogen energy is still in the preliminary stage, subsidies for hydrogen energy are more necessary to formulate different subsidy measures according to different products and enterprise sizes, and the current subsidy policies should not ignore these heterogeneities.

3.2 Innovation system building

The national development and reform commission and national energy bureau released the revolutionary energy technology innovation action plan (2016-2030). It "explicitly mentioned, by 2030, to build sound energy technology innovation system that compatible with national conditions. The improvement of our capacity for independent energy innovation, and make energy technology reached the international advanced level, will help coordinate sustainable development of energy industry in China, and let China enter the world's energy technology powerhouse. The action plan pointed out that although China's energy science and technology level has made great progress and significantly improved, but compared with the world's power in energy science and technology and leading the energy revolution requirements, there is still a big gap.

In order to cope with the competition of intellectual property in the current international market, this paper argues that the government should play its guiding role. First, it is necessary to strengthen enterprises' awareness of efficient patent protection, enhance the intensity of participating in the international market competition, do a good job in international patent application and distribution, and actively seize the commanding heights of scientific and technological innovation. The second is to encourage and strengthen the cooperation between universities and enterprises with traditional technology advantages, combine the advantages of all parties, promote the research, development and application of core technologies, and maximize the market value of technological innovation achievements through patent licensing, patent transfer and other forms. The practical measures of realizing these two points should be to give play to the promotion role of government mechanism, regard technological innovation as the foundation of national progress, establish a complete reward mechanism for researchers who make major breakthroughs in core technology, and establish certification mechanism for relevant enterprises. In addition, at the level of national investment in scientific research, overseas talents should be introduced, and scientific research institutions in regions with superior hydrogen energy endowment should be cultivated.

3.3 Industry standardization construction and social education

The safety of hydrogen energy is the focus of the society. The international and domestic application practice has proved that the safety of hydrogen energy can be guaranteed as long as hydrogen energy is produced, stored and used in accordance with standards and regulations. In terms of the hydrogen and fuel cell standard system construction, the hydrogen standardization technical committee (SAC/TC 309) and the national fuel cells and fluid flow battery standardization technical committee (SAC/TC 342) so far, a total of 26 hydrogen related national standards, fuel cell related national standards, 41, Chinese hydrogen energy and fuel cells preliminarily established national standard system framework. It is undeniable that in the process of promoting the hydrogen energy industry, relevant standards in the field of hydrogen energy in China still need to be further improved, and there are still some uncoordinated and unmatched problems. The degree of integration with international standards also needs to be improved. Therefore, it is necessary to further strengthen the standardization work to support the development of hydrogen energy industry and give full play to the supporting and leading role of the standard to the industrial development. We suggest that the state increase investment in hydrogen energy safety, promote hydrogen energy safety based on the study of the causes, mechanisms and consequences of hydrogen accidents, as well as the methods of accident prediction, early warning and control, and formulate reasonable safety technical specifications and standards. Based on the concept of intrinsic safety, equipment/facilities should be designed to establish a sound quantitative assessment system for hydrogen energy utilization risk, prevent and reduce hydrogen energy safety accidents, and promote the healthy and safe development of hydrogen energy industry.

In addition, the public's acceptance of hydrogen energy needs more scientific popularization and education, in which the efforts of the government and scientific research institutions are needed. Although current concerns about safety are normal, there is a need to prevent social panic and opposition. Let the public more fully understand the characteristics of hydrogen energy, production and use of the correct way to avoid possible safety problems. Just as coal and natural gas were just beginning to become common, governments should play an important role as educators. In addition, as the Alanne [7] study argues, system-level approaches should be integrated into hydrogen education, not just the teaching of chemical reactions or hydrogen technology. Both radical and traditional problem sets are needed to cultivate students' creative thinking and keep realistic.

References

1. Guo, L., & Jin, H. Boiling coal in water: hydrogen production and power generation system with zero net co2 emission based on coal and supercritical water gasification. International Journal of Hydrogen Energy, **38(29)**, 12953-12967, (2013)
2. Chen, J., Fan, Y., Zhao, X., Jiaqiang, E., Xu, W., Zhang, F., ... & Liu, S. Experimental investigation on gasification characteristic of food waste using supercritical water for combustible gas production: Exploring the way to complete gasification. Fuel, **263**, 116735, (2020)
3. Yan, X. L., & Hino, R. *Nuclear hydrogen production handbook* (CRC press, 2016).
4. Eriksson, E. L. V., & Gray, E. M. Optimization and integration of hybrid renewable energy hydrogen fuel cell energy systems–A critical review. Applied energy, **202**, 348-364, (2017)
5. Klumpp, F. Comparison of pumped hydro, hydrogen storage and compressed air energy storage for integrating high shares of renewable energies—potential, cost-comparison and ranking. Journal of Energy storage, **8**, 119-128, (2016)
6. Gis, W., & Schaap, G. Hydrogenation of road transport on the example of Sweden and Poland. IOP Conference Series: Materials Science and Engineering, **421(4)**, 042024, (2018)
7. Alanne, K. A survey of finnish energy engineering students' knowledge and perception of hydrogen technology, International Journal of Hydrogen Energy, **43(22)**, 10205-10214, (2018)
8. Tian, M., Yuen, H. C., Yan, S. R., & Huang, W. L. The multiple selections of fostering applications of hydrogen energy by integrating economic and industrial evaluation of different regions. International Journal of Hydrogen, **44(56)**, 29390-29398, (2019)

Application research progress of hot isostatic pressing technology in nickel-based single crystal superalloy

*Jun Wang**, *Zhaojun Jiang,*

Wuxi Institute of Technology, Wuxi, 214121, R.P. China

Abstract. Since the introduction of hot isostatic pressing (HIP) technology, it has been widely used in powder metallurgy forming and improving the densification of castings. With the increasing demand for the performance of nickel-based single crystal superalloy, more and more scholars started research on the application of hot isostatic pressing in nickel-based single crystal superalloy. This article summarizes the current research progress in the application of hot isostatic pressing to the high temperature of nickel-based single crystals, explains the elimination effect of hot isostatic pressing on porosity and other pore defects in nickel-based single crystal superalloy, analyzes the pore healing mechanism, and the changes in mechanical properties such as tensile, endurance, and fatigue of the alloy after applying hot isostatic pressing are shown. Finally, the future development of China's hot isostatic pressing technology is prospected.

1 Introduction

Nickel-based single crystal superalloy has become one of the most important materials for electric energy gas turbines due to their excellent fatigue and creep resistance at high temperatures and excellent resistance to oxidation and corrosion [1-3]. Compared with other cast superalloy, single crystal superalloy produced by investment casting have little porosity tendency [4], but they still inevitably have defects such as porosity and segregation. Because of the dendrite growth mode during the solidification of alloy crystal, the dendrite region gradually becomes the final solidification region, and the metallic liquid cannot enter the dendrite region to shrink, resulting in the formation of micropores between dendrites. When the alloy is subject to solution heat treatment, defects such as solution micropores will also be formed due to the Kirkendall effect [5-6], the diameter of these pores is about 10-30 μm, and the volume fraction is a few tenths of a percent [7]. During the alloy service life, void defects will become the source of cracks and cause the mechanical properties of the alloy to decline, which seriously reduces the service life of the material [8].

Hot isostatic pressing technology is mainly used to heat metal or ceramic parts while applying the same pressure in all directions, so that the parts can be sintered and densified. The use of hot isostatic pressing technology can not only significantly eliminate defects in the alloy but also improve the internal microstructure and mechanical properties of the

* Corresponding author:boher@126.com

alloy, and improve the reliability of the alloy [9]. At present, the main application fields of HIP technology are densification treatment of castings, powder metallurgy and bonding fixation of materials, accounting for 54%, 43% and 3% respectively. As the HIP technology has unique advantages in promoting the compactness of parts, eliminating void defects, reducing segregation and improving the mechanical properties of the alloy, it has gradually become the standard post-treatment process for turbine blades made of nickel base superalloy [10]. At present, the research of hot isostatic pressing (HIP) technology for the application of electrical energy superalloy in foreign countries has been relatively mature, and the HIP technology has been applied in the engineering of electrical energy superalloy. However, the research and application of HIP technology in the field of Single Crystal Superalloy for gas turbine electrical energy sources in China is not much [11].

In order to eliminate the micro hole defects in gas turbine power materials, it is necessary to study the microstructure changes and pore healing of nickel base single crystal superalloy under different hot isostatic pressure parameters, understand the mechanism of pore healing, and avoid other defects in the alloy caused by hot isostatic pressure. This paper reviews the recent research and application of hot isostatic pressing technology in nickel-based single crystal superalloy for gas turbine power energy materials, analyzes hot research directions of hot isostatic pressing, and points out the development prospect of hot isostatic pressing technology in the field of nickel base single crystal superalloy.

2 Effect of HIP on densification of nickel base single crystal superalloy

2.1 Effect of HIP parameters on porosity of nickel base single crystal superalloy

The main process parameters involved in hot isostatic pressing include temperature, pressure and holding time. Different process parameters play different roles in promoting hole closure and improving the compactness of the workpiece. Taking temperature as an example, when the HIP temperature is higher than the solution temperature of γ' phase, the nickel base single crystal superalloy can be softened in the case of single phase, and the plasticity can be significantly improved, which is conducive to the closure of pores. However, when the temperature is close to the solidus temperature of the alloy, there is a risk of recrystallization and initial melting inside the material, which is strictly prohibited in engineering applications [12-13]. Therefore, in order to control the cost and ensure the quality of the parts, it is necessary to study the control and influence mechanism of different parameters of HIP treatment on the microstructure and properties of the alloy in detail, and select the optimal matching scheme of process parameters, so as to obtain the alloy material with significantly improved comprehensive performance and give full play to the advantages of HIP technology.

Cao [14] studied the effect of different hot isostatic pressing temperature on the internal micro porosity of DD10 single crystal superalloy. Figure 1 shows the optical structure of DD10 single crystal superalloy as cast and treated by different hot isostatic pressing schemes. The results show that the micro porosity in the as cast alloy can be completely eliminated after being treated by isostatic pressing at 1290 ° C and 150MPa. When the pressure is constant and the temperature is raised to 1300~1340 ° C, the micro porosity in DD10 single crystal superalloy can also be completely closed. However, after 1340 ° C, there is primary melting in the alloy. Therefore, the hot isostatic pressure temperature should not exceed 1340 ° C under 150MPa to avoid the initial melting of DD10 single crystal superalloy.

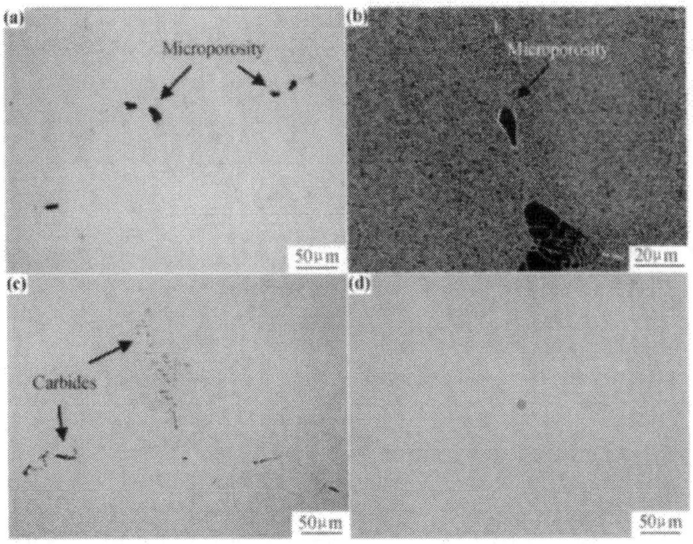

Fig.1. Distribution and morphology of micro porosity in single crystal superalloy DD10 at different conditions.

(a) As cast structure (b) as cast micro porosity SEM morphology

(c) HIP-1290°C/150MPa optical structure (d) HIP-1340°C/150MPa optical structure

Inmaculada Lopez-Galilea [15] carried out a hot isostatic pressing process study on the single crystal superalloy ERBO-1 at different temperatures, pressure and holding time. The as cast sample (ERBO-1A) and the fully solution and aged sample (ERBO-1C) were selected for different hot isostatic pressing studies. The hot isostatic pressing process scheme shown in Table 1 was designed to study the effects of temperature and holding time on the pore closure of the two samples under the determined pressure conditions. Figure 2 shows the material porosity corresponding to different experimental results. It can be seen that the application of hot isostatic pressure has a significant effect on the reduction of material porosity, and the effect is more significant with increasing temperature. Regardless of the state of the sample, when the hot isostatic pressing temperature is higher than the γ 'phase dissolution temperature (about 1260 ° C), the sample pores are almost completely eliminated after the hot isostatic pressing regardless of the holding time of 3h or 5h.

Table 1. Hot isostatic pressure scheme under determined pressure.

Temperatu re (°C)	holding time (h)	Pressure (MPa)
1100,1200, 1220,1320	3	200
1200,1220, 1300,1320	5	200

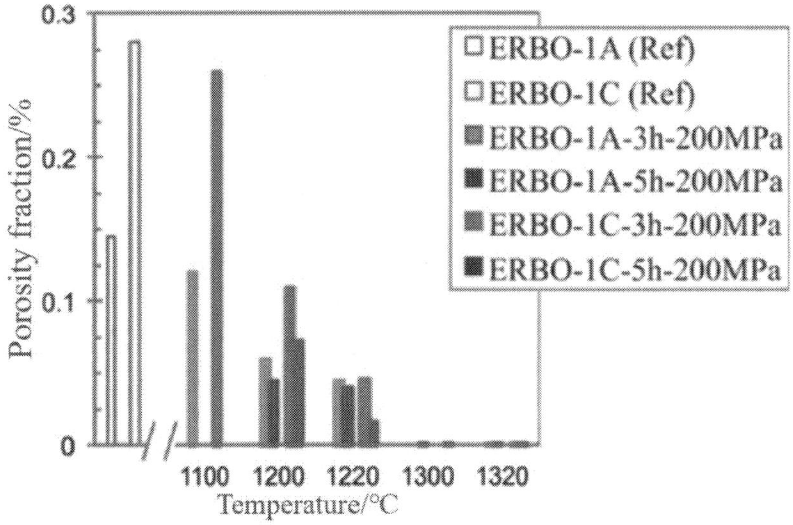

Fig. 2 Porosity of as-cast ERBO-1A and hardened ERBO-1C samples under different hot isostatic pressing schemes.

The hot isostatic pressing process scheme shown in Table 2 is designed to study the effect of temperature and pressure on the pore closure of as-cast samples under the condition of determining the holding time. Figure 3 shows the porosity of materials corresponding to different experimental results. It can be seen that under atmospheric pressure, the porosity of the material increases due to the solid solution micro porosity generated by the internal element diffusion under high temperature. When it is lower than the dissolution temperature of γ 'phase, the porosity will indeed decrease with the increase of applied pressure, as shown in the blue column in Figure 3. However, when the temperature is higher than the dissolution temperature of γ 'phase, the pores have been effectively closed at 100MPa, as shown in the green column in Fig. 3, and further increase of pressure has no additional effect on the porosity.

Table 2. Hot isostatic pressure scheme for determination of holding time.

Temperature (°C)	holding time (h)	pressure (MPa)
1220	3	0.1,200,300
1200,1220,1300,1320	3	100,200

 https://doi.org/10.1051/e3sconf/202015501012

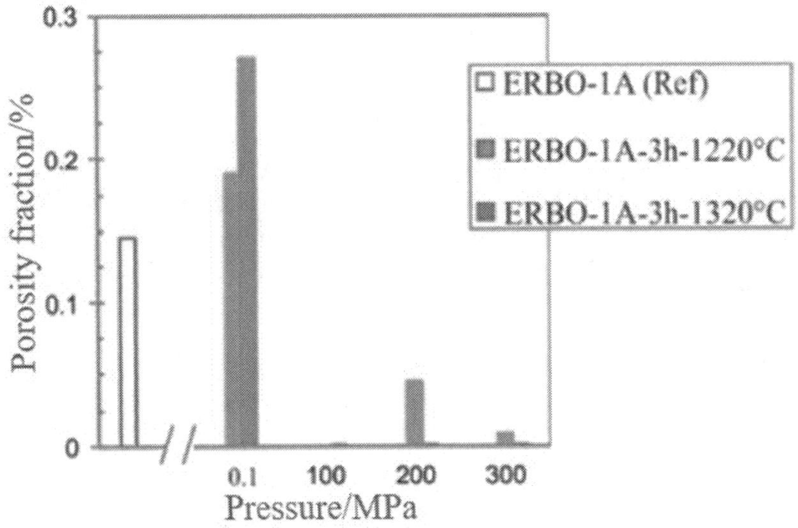

Fig.3. Porosity of as-cast ERBO-1A samples under different hot isostatic pressing schemes.

2.2 Study on hole healing mechanism under hot isostatic pressure

Generally speaking, under high temperature and high pressure, plastic deformation and creep behavior take place in the cavity area of the alloy, and then the surface of the collapse area is bonded under the action of diffusion behavior, so as to improve the densification of the material. Therefore, the internal microstructure of the alloy is optimized, the density is improved, and the composition uniformity is also improved under the action of hot isostatic pressing [16-17].

L. Mujica Roncery [18] thinks that the main closing mechanism of holes is creep deformation. He has carried out a series of studies on the closing mechanism of holes under hot isostatic pressure. Under and above the dissolution temperature of γ' phase, the nickel base single crystal superalloy was treated by hot isostatic pressure respectively. Figure 4 shows the microstructure changes around the holes under the action of hot isostatic pressure. It is found that raft structure appears around the hole under stress when the temperature is lower than the dissolution temperature of γ' phase. When the temperature is higher than the dissolution temperature of the γ 'phase, the dislocation is no longer fixed by the γ' phase due to the dissolution of the γ 'phase in the hot isostatic pressing process, so the deformation of the hole is more rapid.

(a) ~ (c) : lower than the dissolution temperature of γ 'phase ; (d) ~ (e) : higher than the dissolution temperature of γ' phase

Fig. 4. Microstructure of micropore area after hot isostatic pressing.

Alexander Epishin [13] studied the hole healing mechanism of nickel-based single crystal superalloy CMSX-4, and set up a series of interrupt experiments to explore the dynamic mechanism of hole healing. Analyzing the experimental results, the article considers that the hole closure is mainly caused by plastic deformation caused by dislocation slip. At the same time, the theory of crystal plasticity and large strain was used to model the dynamics of hole closure during the hot isostatic pressing process of single crystal superalloy. A physical model of the hole closing process was developed and studied the influence of different parameters of hot isostatic pressing on the dynamic process of hole closure. Figure 5 shows the comparison between the experimental and simulation results and the simulation results also verifies the mechanism of pore closure caused by plastic deformation.

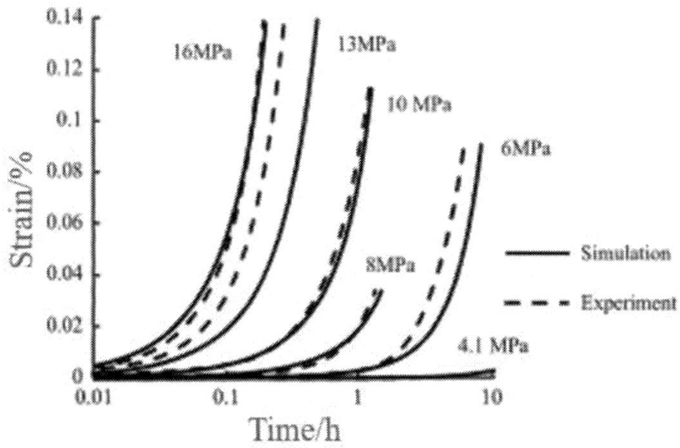

Fig.5. Comparison between experimental creep curve and simulated creep curve.

Zhou et al. [19-20] studied the mechanism of hot isostatic pressure (HIP) on the repair of creep holes, and considered that temperature and pressure are two driving factors of solute diffusion, and their interaction affects the healing performance of creep holes. Based on the characteristics of hole repair and related thermodynamic theory, a hole repair model is proposed. In this model, the stress gradient around the hole is used as the driving force of atom diffusion to study the effect of hot isostatic pressure temperature and pressure on the healing dynamics of creep hole. The relations between the healing time and temperature and pressure was established (Fig. 6), thus providing a basis for the selection of HIP parameters.

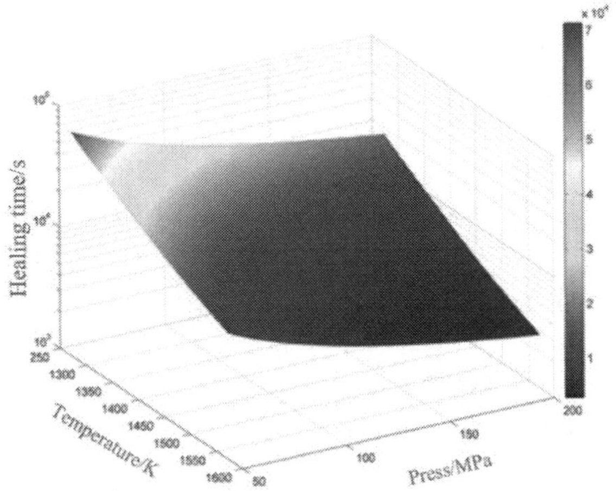

Fig. 6. Simulation of relationsHIP between time required for hole repair and HIP temperature and pressure.

3 Effect of HIP on mechanical properties of nickel-based single crystal superalloy

3.1 Effect of hot isostatic pressing on tensile properties of nickel-based single crystal superalloy

Lan [21] et al. studied the change in tensile properties of single crystal superalloy under hot isostatic pressing. The hot isostatic pressing scheme shown in Figure 7 was used to perform hot isostatic pressing on the fully heat treated second-generation single crystal superalloy CMSX-4. Tensile tests were carried out at 980 °C on two groups of specimens treated with HIP or not, and the test results are shown in Figure 8. The study found that the average elongation at break of the samples without HIP treatment was 25.7%, and the average elongation at break of the samples after HIP treatment increased significantly to 31.3%, while the yield strength remained unchanged. By analyzing the fracture morphology, it can be concluded that HIP treatment can reduce the size and number of micropores in the alloy, thus inhibiting the crack initiation and propagation and improving the tensile properties of the alloy.

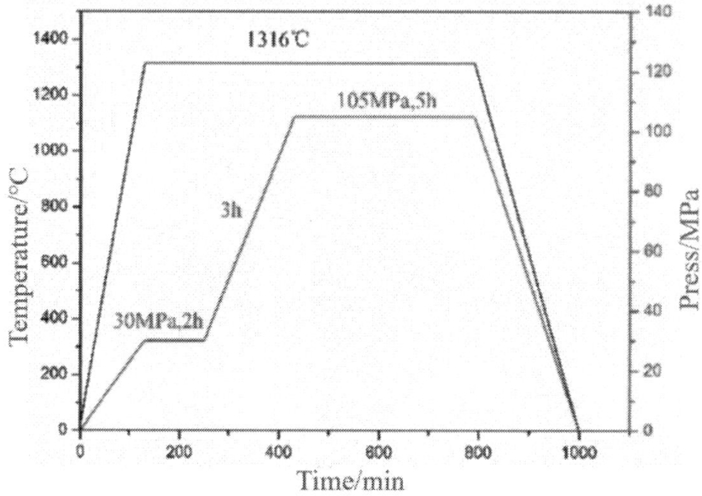

Fig.7. Hot isostatic pressing scheme.

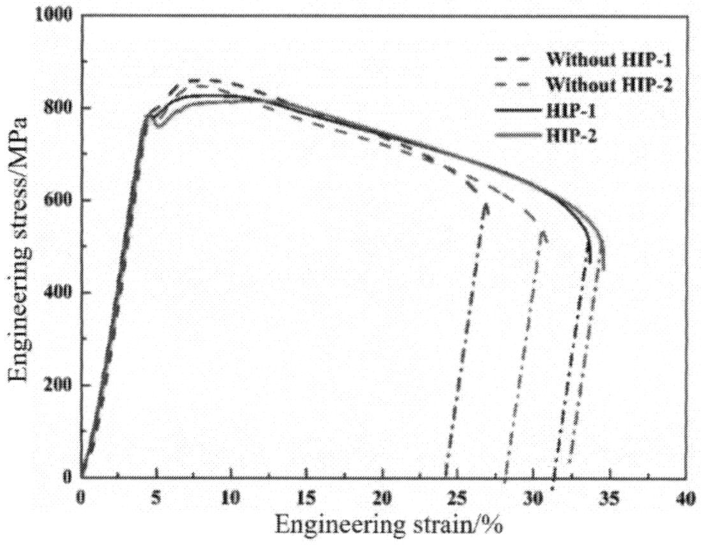

Fig.8. Engineering stress-strain curve of samples with and without HIP treatment at 980 °C.

3.2 Effect of hot isostatic pressing on the durability of nickel base single crystal superalloy

Chang [22] treated the CMSX-4 alloy by hot isostatic pressing, and analyzed the microstructure and mechanical properties of the alloy. The research shows that the internal porosity of the alloy is not only significantly reduced after the hot isostatic pressing treatment, but also the endurance life increased from 72.7h to 134.28h at 950 °C/355MPa. The experimental results are shown in Figure 9.

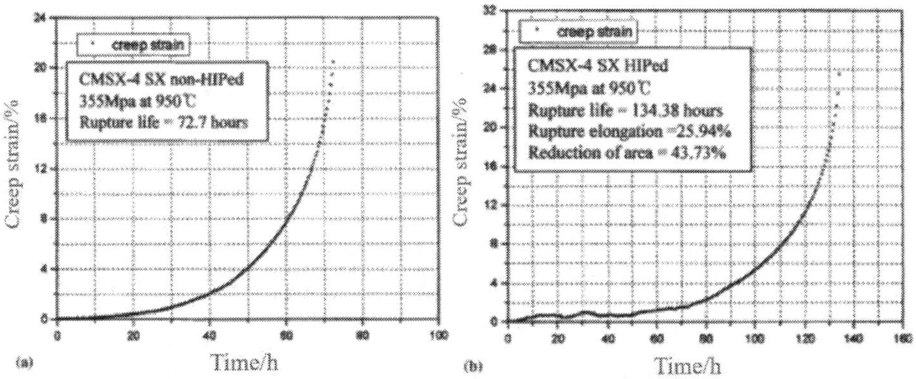

Fig.9. Stress-strain curve of CMSX-4 alloy at 950°C/355MPa.

(a) No HIP treatment; (b) HIP treatment

Shi Zhenxue et al. [23] subjected the as-cast nickel-based single crystal superalloy samples to hot isostatic pressing at 1180°C/150MPa. In order to compare the endurance properties of samples under different conditions, standard heat treatment is carried out for both the samples after HIP treatment and as cast samples, and the experimental results are shown in Table 3. It can be seen that the alloy's durable life is increased after hot isostatic pressing under the conditions of 760°C/800MPa and 980°C/250MPa, while the alloy's durable life has not changed significantly under the condition of 1100°C/140MPa.

Table 3. Effect of hot isostatic pressing on the durability of alloy under different conditions

test conditions	Treatment method	t/h
760°C/800MPa	heat treatment	353.5
	HIP+heat treatment	462.1
980°C/250MPa	heat treatment	225.2
	HIP+heat treatment	254.6
1100°C/140MPa	heat treatment	83.4
	HIP+heat treatment	85.1

4 Summary

With the rapid development of nickel base single crystal superalloy and the sharp increase of domestic demand for nickel base single crystal superalloy products, people are eager to improve the temperature bearing capacity of the alloy and its comprehensive performance at high temperature, so as to develop advanced aircraft engines and gas turbines with higher thrust weight ratio and fuel efficiency. However, pores and other defects that cannot be avoided during the metal casting process restrict the development of the alloy to a certain extent, endanger the mechanical properties of the alloy and reduce the service life of the alloy. The hot isostatic pressing technology, as an effective method to eliminate the defects in the alloy, has been deeply studied by domestic and foreign scholars, and has been successfully applied in engineering. With the advent of hot isostatic pressing equipment with rapid cooling function, scholars have begun to study the combination of heat treatment process and hot isostatic pressing process to treat alloy products, so as to improve the overall performance of the alloy and greatly reduce the cost. At the same time, the research of HIP technology in the field of material repair, the combination of HIP and additive manufacturing and other advanced technologies was also carried out.

At present, the application research of China's hot isostatic pressing technology in the field of nickel-based single crystal superalloy still has a certain gap with foreign developed countries. The blockade of related foreign technologies and the lack of advanced hot isostatic pressing equipment restrict the engineering application of our country's hot isostatic pressing technology in the field of nickel-based single crystal superalloy. Most of the current scientific research work is limited to the characterization of microstructure and material properties while lacking in mechanism analysis and engineering application research. Therefore, the application of hot isostatic pressing technology in the field of nickel base single crystal superalloy still needs the active participation of more scientists. With the rapid development of aeroengine technology in China, the demand for turbine blades of nickel base single crystal superalloy is also increasing. It is believed that the application of hot isostatic pressing technology in the field of nickel base single crystal superalloy will be developed rapidly in the near future.

References

1. Z. Q. Hu, L. R. Liu, T. Jin, & X. F. Sun, Aeroengine, 31 (3), 1-7 (2005).
2. R. Z. Chen, Journal of Materials Engineering, (8): 3-12 (1995).
3. R. C. Reed, The Superalloy: Fundamental and Applications, 1 (2006).
4. J. R. Li, J. C. Xiong, D Z. Tang, et al. National Defense Industry Press, 175 (2012).
5. T. Link, S. Zabler, A. Epishin, A. Haibei, M. Bansal, MSEA, 425: 47–54 (2006).
6. B.S. Bokstein, A. Epishin, T. Link, V.A. Esin, A.O. Rodin, I.L. Svetlov, Scripta Materialia, 57: 801–804 (2007).
7. A. Epishin, T. Link, U. Brückner, P. D. Portella, Materialsfor advanced power engineering, 217-226 (2002).
8. J. Zhang, L. H. Lou, AcatMetallurgicaSinica, 54 (11): 1637-1652 (2008).
9. H. Hu, AcatMetallurgicaSinica, 38 (11): 1699-1202 (2002).
10. A. Weddeling, W. Theisen, Metal Powder Report, 72 (5): 345-348 (2017).
11. M. Han, Y. S. Luo, Journal of Materials Engineering, (8): 40-43 (2008).
12. R. Bürgel, P. D. Portella, J. Preuhs, Superalloy. Champion, USA: TMS, 229–238 (2000).
13. A. Epishin, B. Fedelich, T. Link, et al. Materials Science & Engineering A, 586: 342-349 (2013).
14. L. M. Cao, L. J. Liu, J. Y. Chen, et al. Journal of Materials Engineering, (6): 1-4 (2013).
15. A. Epishin, T. Link , B. Fedelich, et al. MATEC Web of Conferences. Vol. 14. EDP Sciences, 2014.
16. J. C. Chang, C. Choi, J. C. Kim, et al. Journal of Materials Engineering and Performance, 12 (4): 420-425 (2003).
17. R. F. Huang, W. H. Yu, Y. J. Tang, et al. Materials of Mechanical Engineering, (3): 50-53 (1988).
18. M. L. Roncery, I. Lopez-Galilea, B. Ruttert, et al. Materials and Design, 97: 544-552 (2016).

19. **Y**. Zhou, S. X. Rao, Z. Zhang, et al. Materials and Design, 49: 25-27 (2013).
20. **Y**. Zhou, Z. Zhang, Q. P. Zhong. et al. Computational Materials Science, 65: 320-323 (2012).
21. J. Lan, W. D. Xuan, Y. Han, et al**.** Journal of Alloys and Compounds, 805: 78-83 (2019).
22. **J**. C. Chang, C. Choi, J. C. Kim, et al. Journal of materials engineering and performance, 12 (4), 420-425 (2003).
23. **Z**. X. Shi, S. Z. Liu, J. C. Xiong, Rare Metal Materialsand Engineering, 44 (9): 2300-2304 (2015).

A DFT study of diesel exhaust NO_x reduction over rare earth CeO_2 catalyst

He Huang [1,*], *Xiao Zhang*[2]

[1] School of Traffic Engineering, Nanjing Institute of Industry Technology, Nanjing 210046, China

[2] Zhenjiang Campus, Army Military Transportation University of PLA, Zhenjiang 212000, China

Abstract. Adsorption of NO on the CeO2(110) surface was investigated using RPBE approach of GGA within the framework of density functional theory (DFT) combined with periodic slab model. Two molecular orientations, N-end and O-end, over various adsorption sites, top, hollow, bridge and O site of CeO2(110) sur-face have been considered. Two molecular orientations under different coverage of CeO2(110) surface also have been considered. The optimized results indicate that the N-end adsorption models are more stable than the O-end ones. So N-end adsorption was more favourable than O-end. NO adsorption on a clean CeO2(110) surface was physisorption, while chemical adsorption occurred in the present of surface oxygen vacancy. Adsorption is stable when coverage was set to 0.25 monolayer. Researched on density of states of free NO molecule and adsorbed NO molecule, the results show that there is an interaction between NO molecule and the substrate . The population analysis indicates that the charges transfer from Ce atoms to NO molecule. The charges transfer of O-end is more than N-end.

Cerium dioxide (CeO_2) is one of the most suitable three-way catalysts (TWCs) for automobile exhaust emission system[1]. It can simultaneously convers hydrocarbons (HCs), CO and NO_x from the exhaust into H_2O, CO_2 and N_2. CeO_2 functions as oxygen buffer due to its great capacity of oxygen storage/release[2-3]. The oxygen vacancy in CeO_2 enables fast oxygen storage when exhaust oxygen concentration is high, and opportune oxygen release when the concentration is low[4-5]. The key reactions for exhaust emission control are CO oxidation and NO_x reduction, and adsorption is the initial step for the reactions. In previous research, periodic density functional theory was implemented to study adsorption pattern of CO on CeO_2(110) and CeO_2(111) surfaces, which indicated the oxidation of CO was due to the lattice oxygen on CeO_2 surface.

There have been many studies about NO adsorption on CeO_2 surface. Ozeki et al. [6]found that high adsorption capacity of CeO_2 to NO ((4.9 molecules /nm^2) could be achieved under 303K. Nearly 40% NO chemically bind to CeO_2 as NO_2^-. Based on IR frequency analysis, it was speculated that Ce^{4+} converted to Ce^{3+} due to NO. Daturi et al. [7] analyzed infrared absorption spectra and mass spectrometry and proved that total oxygen exchange of CeO_2 was strongly correlated with NO activity and oxygen vacancy

* Corresponding author: 394807515@qq.com

concentration on CeO_2 surface. Ceria based catalysts have sufficient oxygen vacancy to accelerate NO decomposition. However, due to limitation of methods, it is unfeasible to reveal the site, configuration, energy and electronic structure of NO adsorption. Recently, designing new materials from atomic and molecular level has been widely used. Combined with experimental methods, it can greatly reduce dependency on instruments and chemical reagents and theoretically study underlying mechanisms of catalytic reactions. Therefore, it is a suitable ideology to find selective, active and anti-toxic catalyzer. In this paper, the interaction between NO and CeO_2 was discussed by theoretical calculation.

The NO adsorption pattern on $CeO_2(110)$ surface was systemically studied under density functional theory framework[8-13]. Firstly, models of NO adsorption by N terminal and O terminal on various adsorption sites on $CeO_2(110)$ surface were calculated. Secondly, stability and energy of adsorption sites of NO on $CeO_2(110)$ surface with oxygen vacancy under different coverage were investigated and charge transformation was revealed by state density and layout analysis.

1 Model and Calculation

The Pure CeO_2 is cubic fluorite crystal with face-centered cubic atomic arrangement in unit cell. Each Ce atom is surrounded by eight oxygen atoms, with space group of Fm3m and lattice parameters of a=b=c=5.411Å , $\alpha=\beta=\gamma=90°$[14]. The DFT calculation suggests $CeO_2(110)$ surface has the lowest formation energy of oxygen vacancy to generate O_v, compared to $CeO_2(100)$ and $CeO_2(111)$ [15-16]. Therefore, it can provide more surface reactive oxygen species highly active in catalytic reactions[17]. In this research, $CeO_2(110)$-2×2 and $CeO_2(110)$-3×3 supercells were chosen for NO adsorption by N terminal and O terminal. Figure 1. Shows the models of $CeO_2(110)$-2×2 and $CeO_2(110)$-3×3 supercells.

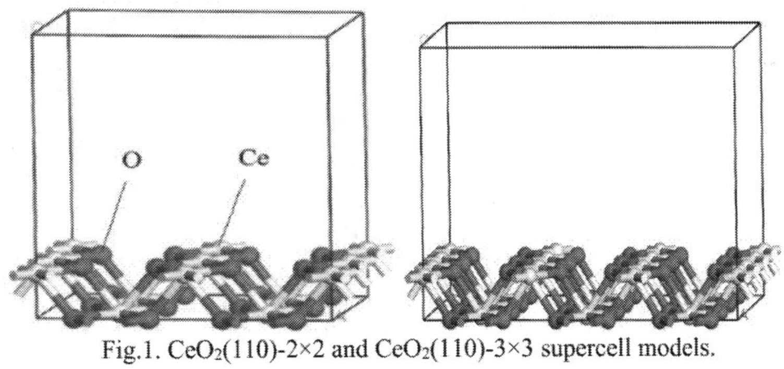

Fig.1. $CeO_2(110)$-2×2 and $CeO_2(110)$-3×3 supercell models.

The NO adsorption on $CeO_2(110)$ surface was simulated by combined method of periodic slab model and GGA-RPBE of density functional theory. All calculation was carried out by Dmol^3 software kit[18-21]. For the calculation of geometry structure optimization and transition state, density functional theory has been proved to be a reliable tool[22]. Core electrons of CeO_2 atom was substituted by effective core potential (ECP) during calculation. The wave functions were expanded by double number polarization function (DNP). All electron basis set was implemented for N and O atoms. Cutoff energy was set as 380eV. Brillouin zone was sampled at k point with Monkhorst-Pack method[23]. The set of accuracy in energy calculation and structure optimization was as following: (1) Self-consistent field energy convergency was 1.0×10^{-4}eV (2) Maximum force was $0.2eV\cdot nm^{-1}$. Converging energy from optimization was less than 1.0×10^{-3}eV. Through geometric optimization, bond length of N-O in free NO molecule was 1.169 Å, which was similar to the result from experiments, 1.151Å[24]. 10 Å vacuum was reserved between two layers to erase potential interactions.

Formation energy of oxygen vacancy on $CeO_2(110)$ surface was represented by:

$$E_V = E_{VCeO_2} + 1/2E_{O_2} - E_{CeO_2} \quad (1)$$

Where E_{VCeO2} was the energy of single oxygen vacancy on $CeO_2(110)$ surface, E_{O2} was the energy of gas oxygen molecule, and E_{CeO2} was the surface energy of clean CeO_2.

The adsorption energy on clean $CeO_2(110)$ surface was calculated as the change of total energy of all substances during the adsorption:

$$E_{ads} = E_{NO} + E_{substrate} - E_{(NO\ +substrate)} \quad (2)$$

Where E_{ads} was adsorption energy, E_{NO} was energy of NO before adsorption, $E_{substrate}$ was energy of substrate before adsorption. $E_{(NO+substrate)}$ was the energy of the whole system after adsorption. Minus value of E_{ads} indicated NO adsorption release heat, vice versa.

The adsorption energy on $CeO_2(110)$ surface with oxygen vacancy was defined as:

$$E_{ads} = -(E_{NO/VCeO_2} - E_{NO} - E_{VCeO_2}) \quad (3)$$

Where $E_{NO}/VCeO_2$was the energy of NO and CeO_2 surface interaction system.

2 Results and discussion

2.1 The calculation of NO molecule

Firstly, the bond length and binding energy of NO molecule were calculated by different methods to verify the reliability of the method selected in this study. The results are shown in table 1. GGA-ROBE method generated closest results to experimental results. So this method was implemented in following steps.

Table 1. The bond length and binding energy of free NO molecule.

Method	R_{N-O} (Å)	BE (kJ/mol)
LDA-PWC	1.154	819.0
LDA-VWN	1.154	818.6
GGA-PW91	1.163	726.5
GGA-BP	1.165	705.7
GGA-PBE	1.164	727.1
GGA-BLYP	1.169	698.0
GGA-VWN-BP	1.169	698.0
GGA-RPBE	1.169	687.4
Exp	1.151	627.2[25]

2.2 The NO adsorption on clean $CeO_2(110)$ surface and the surface with oxygen vacancy

2.2.1 The comparison of NO adsorption at various sites on CeO2(110) surface

In order to investigate the adsorption mode of NO on the surface of $CeO_2(110)$, the geometric configuration parameters and adsorption energy of NO on the surface of $CeO_2(110)$ with oxygen holes and clean were calculated. According to the projection position of center of NO on $CeO_2(110)$ surface, four adsorption sites were considered, including Top site, Hollow site, Bridge site and O site. Figure 2 illustrates the four sites.

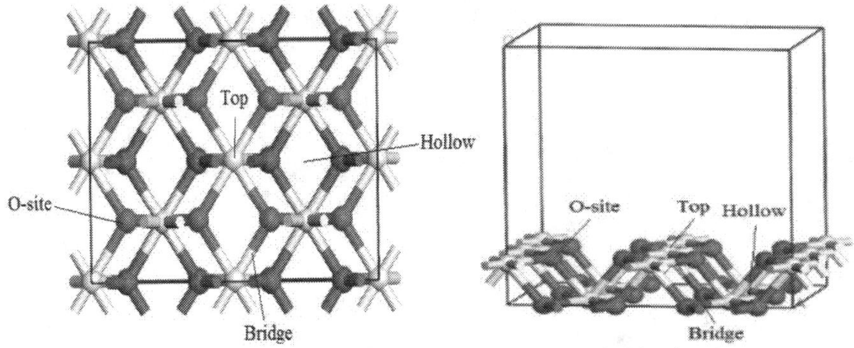

Fig.2. The slab model of $CeO_2(110)$-2×2(top and side views).

Table 2 indicates the adsorption energy under coverage of 0.25mL. On clean $CeO_2(110)$ surface, the adsorption energy was less than 40kJ/mol and the adsorption was physical. N terminal adsorption was more stable than O terminal at each site. The energy of adsorption by N terminal was O > top > hollow > bridge. The energy was highest with N terminal at O site but only reached to 39.6 kJ/mol. The distance between N and O atoms was 2.55 Å. These results were consistent with the results from Yang et al[26], which suggests minimum NO adsorption occurred on $CeO_2(110)$ surface with complete stoichiometric ratio. The increasing N-O bond length by adsorption suggests the adsorption weakened the NO bond, which is beneficial to further decomposition of NO. In this case, O terminal adsorption was more effective than N terminal adsorption.

Table 2. The predicted geometrical parameter and adsorption energies for NO on stoichiometry $CeO_2(110)$ surface.

Adsorption site	Top site		Hollow site		Bridge site		O site	
	N	O	N	O	N	O	N	O
R_{N-O}(Å)	1.188	1.191	1.186	1.203	1.203	1.204	1.199	1.209
Eads/(kJ/mol)	17.4	8.55	15.3	6.2	13.3	4.8	39.6	17.66

2.2.2 NO adsorption on $O_v/CeO_2(110)$ surface

Based on the above analysis, NO has the strongest adsorption at the n-terminal at the oxygen level, so we discuss the n-terminal adsorption at the oxygen level on the surface of $O_v/CeO_2(110)$. Figure 3(a) is the structure of oxygen vacancy on $CeO_2(110)$ surface. By calculation, the formation energy of oxygen vacancy was 180.4kJ/mol which was similar to the result from references[27]. Figure 3(b) was NO adsorption configuration on oxygen vacancy. After adsorption, NO bond length was 1.149Å, shorter than the bond length in free NO molecule. The oxygen significantly increased adsorption energy which were

135.1kJ/mol with N terminal and 80.1kJ/mol with O terminal. Additionally, there was certain interaction between NO and surface of oxygen vacancy. Therefore, NO adsorption will be more effective with N terminal on $CeO_2(110)$ surface with oxygen vacancy.

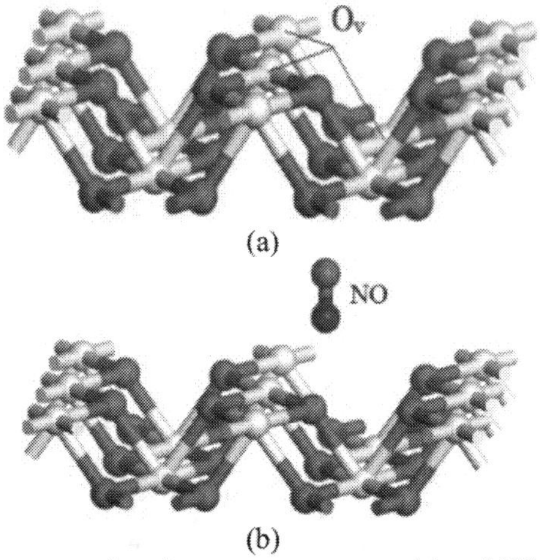

Fig.3. Optimized structures of surface oxygen vacancy (a) and NO adsorbed on the oxygen vacancy(b) O_v:oxygen vacancy;bond length in Å.

2.3 The active of NO adsorption on $O_v/CeO_2(110)$ under different coverage

Geometric optimization was implemented to configuration under coverage of 0.25, 0.33, 0.50, 0.67mL and 1.00mL. Table 3 shows the results. The coverage was defined as ratio of number of adsorbed NO molecules and number of Ce atoms on the surface. For instance, when one NO adsorbed by $CeO_2(110)$-2×2 surface, the coverage was 0.25mL. From table 3, adsorption energy decreased with increasing coverage. However, the energy increased when coverage was 1.00mL. The reason was that due to decreasing distance between NO molecules the interaction of NO enhanced with increasing coverage. The decrease of systemic energy after adsorption caused increasing adsorption energy. The calculation of bond length suggested that the NO bond length increased after adsorption and reached highest under coverage of 0.25mL, which was most suitable for further reaction of NO.

Table 3. The predicted geometrical parameter and adsorption energies for NO over $O_v/CeO_2(110)$ surface.

Coverage(mL)	0.25	0.33	0.50	0.67	1.00
R_{N-O}(Å)	1.199	1.196	1.187	1.185	1.182
E_{ads}/(kJ/mol)	135.1	129.3	107.5	106.8	115.2

To deeply understand the mechanisms of NO adsorption on $O_v/CeO_2(110)$, state density was calculated before and after adsorption. The results are shown in Table 4. There was a peek of state density of free NO at Femi. Compared to Figure 4(a), there was obvious changes of state density of free NO in Figure 4(b)-(f), which suggested that NO interacted with substrate during adsorption and energy band shifted to low energy level. Based on these results, it was speculated that the formation of adsorption bond was caused by the interaction between d orbit in Ce atom and inverse π orbit in NO molecule. Since d orbit with full electrons while inverse π orbit contains electron vacancy, the electron transferred from d orbit to inverse π orbit, causing elongation of N-O bond.

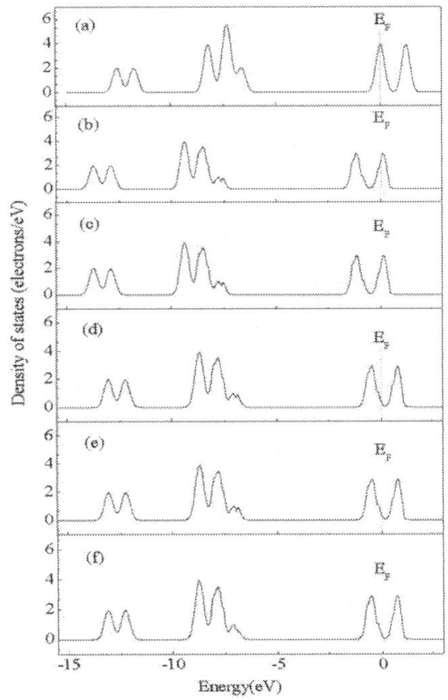

Fig.4. The density of states (DOS) of free NO molecule and adsorbed NO molecule under different coverages.
(a) Free NO, (b) 0.25 ML, (c) 0.33 ML,
(d) 0.50 ML, (e) 0.67 ML, (f) 1.00 ML

2.4 NO molecular layout analysis

The NO molecular layout under coverage of 0.25mL was further analyzed. Table 4 shows the Mulliken charge density of N and O atoms when NO was adsorbed on $O_v/CeO_2(110)$ with N and O terminal. The charges of N and O atoms both changed, and O terminal shifted more than N terminal. It was because O atom contained greater electronegativity and more tended to attract electrons. Negative charge appeared on NO after adsorption which indicated electrons transferred form Ce to NO. This was consistent with the results of state density analysis. Based on Table 4, more electrons transferred during O-terminal adsorption. It was caused by that NO approached substrate with O atom containing greater electronegativity. More electron transferred to NO and weakened N-O bond. Hence, greater change of NO bond length occurred during O-terminal adsorption.

Table 4. The Mulliken charges for N and O atoms in NO molecule.

	N-end	O-end	Free NO
q_N	0.019	-0.023	0.050
q_O	-0.124	-0.181	-0.050
q_{NO}	-0.105	-0.204	0.00

3 conclusions

This research calculated the NO adsorption on clean $CeO_2(110)$ surface or the surface with oxygen vacancy, with periodic slab model by RPBE approach of the generalized gradient approximation within the framework of density functional theory. The following conclusions are drawn.

(1) NO adsorption was physical with low adsorption energy on clean $CeO_2(110)$ surface. When oxygen vacancy existed on the surface, adsorption significantly increased and chemical adsorption occurred.

(2) The lengthening N-O bond after adsorption was beneficial to further decomposition of NO.

(3) The results under various coverage indicated that the smaller coverage induced more stable adsorption. This was caused by mutual repulsion between NO molecule. Therefore, the adsorption was most stable with the coverage of 0.25mL.

(4)Layout analysis suggested that charges transferred from Ce to NO in the whole system. More charges transferred in O-terminal adsorption than in N-terminal adsorption.

References

[1] J. Kaspar, R. Fomasiero, M. Graziani. *Catal. Today.* **50** (1999)
[2] G. R. Li, D. L. Qu, Z. L. Wang, et al. *Chemistry Common.* **48** (2009)
[3] B. T. Teng, S. Y. Jiang, L. H. Zhao. *J. Chinese Rare Earth Soc.* **28** (2010)
[4] M. Sugiur, M. Ozawa, A. Suda, et al. *Bull. Chem. Soc. Jpn.* **78** (2005)
[5] H. F. Wang. *J. East China univ. sci. technol.* **15** (2012)
[6] G. Li, K. Kaneko, S. Ozeki. *Langmuir.* **1** (1997)
[7] M. Daturi, N. Bion, J. Saussey, et al. *Phys. Chem.* **3** (2001)
[8] M. Gaus, Q. Cui, M. *Elstner. Wires Comput. Mol. Sci.* **4** (2014)
[9] X. Wang, W. K. Chen, X. L. Xu. *Chinese J. Catal.* **28** (2007)
[10] X. Wang, W. K. Chen, B. Z. Su. *Chinese J. Inorg. Chem.* **23** (2007)
[11] K. Lejaeghere, V. Speybroeck, G. Ost, et al. *Crit. Rev. Solid State Mater. Sci.* **39** (2014)
[12] W. Kohn, A. D. Becke, R. G. Parr. *J. Phys. Chem.* **100** (1996)
[13] G. X. Luo, Z. H. Li, Z. X. Yang. J. Chin. *Rare Earth Soc.* **25** (2007)
[14] R. O. Jones, O. Gunnarsson. *Rev. Mod. Phys.* **61** (1989)
[15] S. Q. Shi, Y. H. Tang, C. Y. Ouyang, et al. *J. Phys. Chem. Solids.* **77** (2010)
[16] M. Nolan, J. E. Fearon, G. W. Watson. *Solid State Ionics.* **777** (2006)
[17] D. O. Scanlon, N. M. Galea, B. J. Morgan, et al. *J. Phys. Chem. C.* **775** (2009)
[18] M. Nolan. *J. Mater. Chem.* **27** (2011)
[19] L. H. Thomas. *Mathematical Proceedings of the Cambridge Philosophical Soc.* **25** (1927)
[20] S. Y. Jiang, B. T. Teng, J. H. Yuan. *Acta Phys. Sin-Ch Ed.* **25** (2009)
[21] N. Michael, G. Sonja, C. Dean. *Surf. Sci.* **57** (2005)
[22] X. T. Thi, C. Stephen, C. Parkerband. *J. Chem. Soc.* **26** (1992)
[23] X. Y. Zong, M. F. Zhao, W. W. Yan. *Chem. Phys. Lett.* **45** (2008)
[24] M. Daturi, N. Bion, J. Saussey, et al. *J. Phys. Chem.* **5** (2001)
[25] G. Herzberg. *New York: Lancaster.* 1946.
[26] X. Y Zong， K. Tom, K. Hermansson. *J. Chem. Phys.* **124** (2006)
[27] D. Adam. *J. Phys. Chem. C.* **112** (2008)

Flow field analysis of multi-stage middle-open centrifugal pump inter-stage flow passage

Xiao Zhang[1,*], *He Huang*[2]

[1]Zhenjiang Campus, Army Military Transportation University of PLA, Zhenjiang 212000, China

[2]School of Traffic Engineering, Nanjing Institute of Industry Technology, Nanjing 210046, China

Abstract. In order to study the flow velocity, static pressure and turbulent kinetic energy distribution of the inter-stage flow passage, the numerical calculation of the inter-stage flow passage of the multistage split centrifugal pump was carried out under the design condition. The results show that the fluid flows along the inter-stage water flow channel, and backflow and vortices are generated at the guide vanes at the end of the bridge, which causes certain energy loss. In this paper, based on the original design, three different improvement schemes are proposed by changing the shape and position of the guide vane for the backflow and vortex generated near the guide vanes. The improved scheme is numerically simulated, and the energy loss values of the four different flow passages are calculated by integration. After comparison and analysis, the second scheme is determined as the best scheme, and the accuracy of simulation is verified by experiments..

With the rapid development of the national economy, the gradual deterioration of natural conditions and the shortage of water resources, the demand for high-efficiency and low-energy multistage middle-open centrifugal Pumps is increasing year by year in various industries. Compared with the traditional multistage centrifugal pumps, it has a larger flow, higher head and power as well as the efficiency [1-3], which meets the needs of China's economic development. At present, there is no mature theory for the design of multistage middle-open centrifugal pumps in China, and there are many problems in the domestic one, the efficiency is much lower than that of imported pumps [4-5]. So how to improve the efficiency is the main research direction of multistage middle-open centrifugal pump. Its internal structure is complicated, and the flow channel between the two stages is seriously distorted. Therefore, by studying the flow field of the multi-stage pump flow channel, analyzing the factors that affect the performance of the pump and improving the structure are important methods to improve the overall performance and efficiency of the pumps [6-8].

In this paper, the flow field of the flow passage between stages of the multistage middle-open centrifugal pump is simulated, the characteristics of the flow field and the flow law are analyzed, and on the basis of the original design, the scheme of changing the shape and

* Corresponding author: 339214603@qq.com

position of the guide vane is put forward, so as to achieve the purpose of improving efficiency, energy saving and emission reduction.

1 Computation model

1.1 Model design

The main design parameters of the multistage middle-open centrifugal pump are Q = $1.6m^3/s$, H = 225m, P = 5000kW, n = 990r/min, efficiency η = 85%, and the working medium is clean water. The pump adopts a horizontal center-open pump shell, with suction at both sides and middle extrusion; In order to balance the axial force to ensure the smooth operation, the impellers are symmetrically distributed from left to right. From both sides to the middle, the impellers are the first stage, the second stage and the last stage, in which the first stage impeller is single suction on both sides, and the last stage is double suction impeller. The overall structure of the pump is shown in Figure 1.

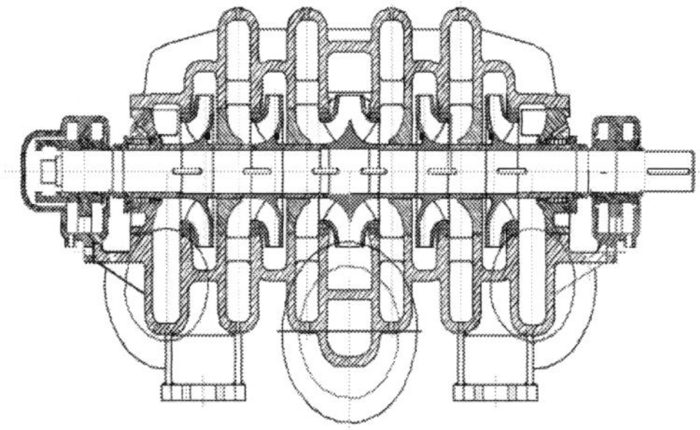

Fig.1. The structure of the Multistage pump.

The multistage pump inter-water flow channel is composed of a double volute and a bridge. The bridge refers to the part connecting the volute and the next stage impeller. The high-speed liquid produced by the impeller enters the 180° symmetrical double volute [9], and then flows through the bridge to the lower impeller. The volute of the inter-stage water flow channel is composed of two single volutes staggered by 180°. This structure produces the same pressure distribution on the two semicircles around the impeller. In theory, the radial forces of the impeller can offset each other. Due to the complicated structure of the inter-stage water flow channel and the difficulty in processing, a rectangular cross section can be used for the volute.

Selection of the radius of the tongue: The liquid flow at the tongue is impacted, which is prone to backflow or vortex. Reasonably selecting the radius of the tongue plays an important role in reducing the hydraulic loss. If the radius of the selected tongue is too small, it is easy to cause cavitation and accompanied by noise and vibration, which will reduce the efficiency of the pump. But if the radius of the tongue is too large, in addition to increasing the radial size of the pump casing, a certain amount of energy is also consumed, so that the efficiency of the pump decreases, so it is necessary to comprehensively consider the appropriate selection of the tongue radius. There is no special theory or empirical formula for bridge design, mainly referring to fluid flow law and empirical design.

Using Pro/E software, the three-dimensional modelling of the multistage middle-open centrifugal pump inter-stage water flow channel is modelled as shown in Figure 2. The double volutes are symmetrically distributed at 180 °. There are two symmetrical guide vanes on the bridge, the main function of which is to introduce the fluid into the lower impeller.

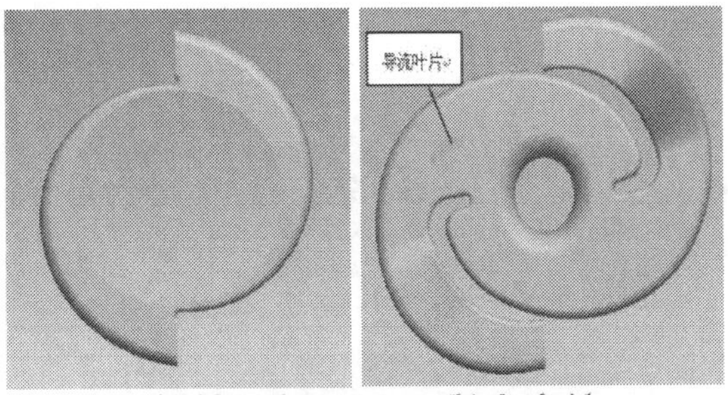

(a) double volute (b) the bridge

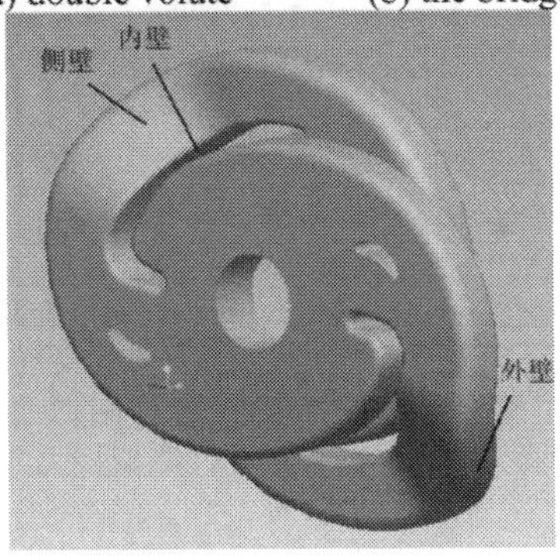

(c) Interstage water passage

Fig.2. Model graph of interstage water passage.

1.2 Mesh generation and independence verification

The structure of flow passage between stages of the multistage middle-open centrifugal pump is complex and twisted seriously. Therefore, unstructured grid with strong adaptability is selected for grid generation[10-11]. The appropriate grid size is defined according to the degree of distortion, and the grid is encrypted in the place with a large degree of distortion and near the wall, and the grid independence is analyzed. For the convenience of calculation, the simulation uses three-stage water bodies with single suction on both sides of the first stage and double suction on the final stage.

The number of grids is approximately 2.2 million, 2.6 million, 3.2 million, 3.8 million, and 4.5 million, and the head of numerical simulation is 245.2m, 245.9m, 246.3m, 246.3m, 246.3m, and the pump efficiency is 78.19%, 78.20%, 78.21%, 78.21%, 78.21%. It can be seen that the calculation results of the five numbers are relatively close. Especially, the efficiency of the three grids above 3.2 million is almost consistent with the head calculation results and stable. Considering the time coordination and calculation accuracy, the number of grids can be 3.8 million, which meets the requirements, so it is selected for research.

1.3 Governing equations

According to the continuity hypothesis of fluid, fluid particles occupy the entire flow space continuously without gaps. The mass of the fluid is neither generated nor disappeared. That is, the increase or decrease of the mass of the fluid inside the control body is equal to the flow into or out of the surface the mass of the fluid, which is expressed by the differential equation as follows:

$$\frac{\partial \rho}{\partial t}+\frac{\partial}{\partial x_i}(\rho u_i)=0 \tag{1}$$

Where ρ is the fluid density and u_i is the velocity component in i direction.
The momentum equation is:

$$\frac{\partial(\rho u_i)}{\partial t}+\frac{\partial(\rho u_i u_j)}{\partial x_j}=-\frac{\partial p}{\partial x_i}+\frac{\partial}{\partial x_j}\left(\mu\frac{\partial u_i}{\partial u_j}\right)+\frac{\partial \tau_{ij}}{\partial x_j} \tag{2}$$

Where $\tau_{ij}=-\rho u_i' u_j'$ belongs to the Reynolds stress terms. In this paper, the standard k-ε [12] model is used for numerical calculation.

1.4 Calculation methods and boundary conditions

The software Flunet is used for calculation and simple algorithm is used. The terms of momentum, turbulent kinetic energy and dissipation rate are in the second-order upwind format, the wall surface adopts the non-slip boundary condition the inlet section is the flow inlet while the outlet section is the static pressure one, and the dynamic-static interface model adopts Frozen rotor. The influence of roughness is considered on the solid wall and the near-wall area is processed using the standard wall surface function [13].

2 Calculation results and analysis

In this paper, the multistage middle-open centrifugal pump is numerically simulated under the design conditions, and the pressure, velocity and turbulent kinetic energy distribution of the inter-stage flow passage are analyzed. The effect of the inter-stage flow passage on the pump performance is emphatically studied, and three different improvement schemes are put forward for the guide vane to optimize.

2.1 Simulation results analysis of inter-stage flow passage

Figure 3 shows the static pressure distribution of the impeller and the double volute section. The figure shows that the static pressure in the central section of the double volute is basically symmetrically distributed at 180°. The main reason for the radial force is the pressure imbalance on the coupling surface between the impeller and the volute, which due to the rotation of the impeller. The static pressure gradually increases and changes evenly from the tongue to the diffusion section. It can be seen from the static pressure distribution of the impeller in the figure that the use of symmetrical double volutes can basically offset the radial force and play a better balance.

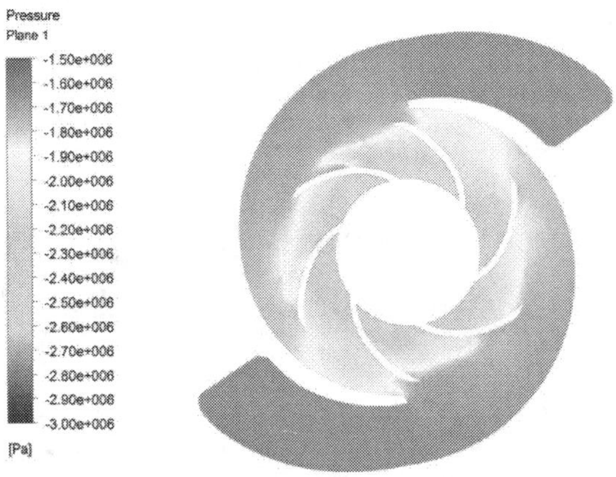

Fig.3. Static pressure profile of the impeller and the double volute.

Figure 4 shows the absolute velocity distribution near the center section and the tongue of the double volute. It can be seen in the figure that the absolute velocity of the fluid gradually decreases from the tongue to the bridge entrance, which also verifies that the volute converts the kinetic energy generated by the impeller into pressure potential energy. In addition, there are several local increase areas with 180 °symmetry of velocity vector near the inner ring surface of double volute, and the increase position is the same as that of static pressure reduction in Figure3. This is because the velocity of the fluid flowing through the blade body increases when the fluid is introduced into the double volute, and a part of the potential energy becomes kinetic energy. The distribution of velocity field near the septum is reasonable and there is no obvious circumfluence and vortex, which shows that the size design of the tongue is reasonable.

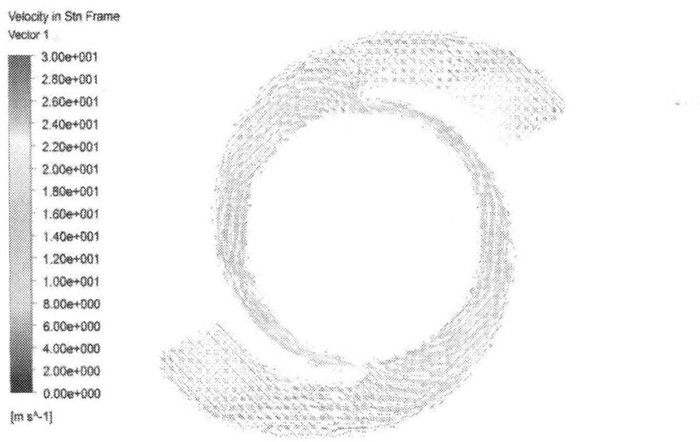

Fig.4. Absolute velocity distribution on the double volute and the volute tongue.

Figure 5 shows the static pressure distribution on the inner and outer walls of the bridge. After the water flow enters the bridge, the speed decreases, and the kinetic energy is further converted into pressure potential energy. The static pressure increases gradually along the flow direction from the bridge inlet, and the static pressure on the inner wall is smaller than that on the outer wall. The pressure fluctuates at the entrance of the bridge, which is caused by the obvious space distortion at the entrance of the bridge, resulting in the inconsistent change range of the flow direction. At the outlet of the inner and outer wall, the local pressure decreases, and the inner pressure gradient is large. This is because the inner flow channel is shorter, the distortion changes are obvious, and the sharp loss of the liquid causes the impact loss on the wall surface to increase, and the potential energy decreases; while the outer flow channel is longer, the transition is better, the impact loss is smaller, and the potential energy is only in a small range Inside has decreased.

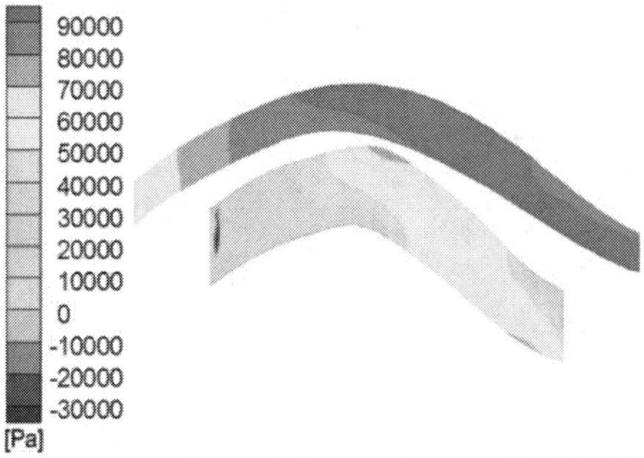

Fig.5. Static pressure profile on the inner and outer wall of the flow bridge.

Figure 6 and Figure 7 show the velocity distribution of the four walls of the bridge. It can be seen from the figures that the speed track at the bridge entrance is basically the same as the contour, and the speed track changes accordingly with the change of the bridge size. The space of the bridge is distorted seriously, the size changes not only in the radial direction, but also in the axial direction.

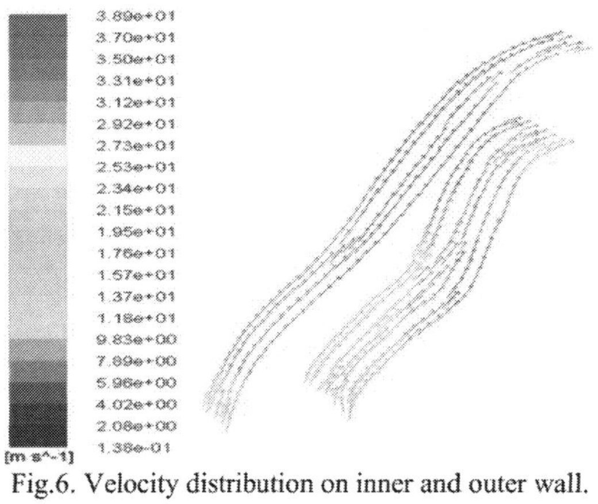

Fig.6. Velocity distribution on inner and outer wall.

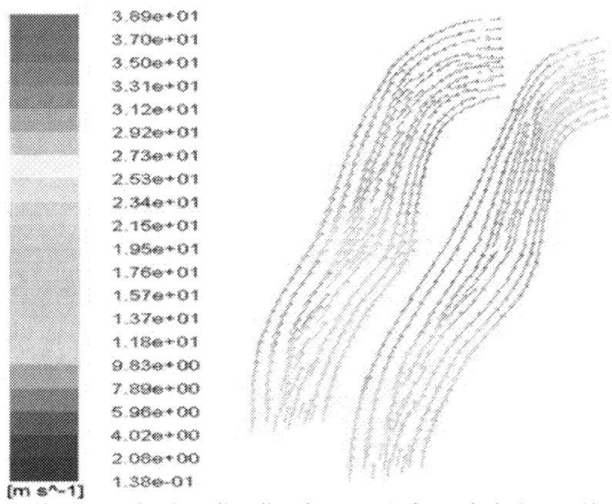

Fig.7. Velocity distribution on left and right wall.

Figure 8 shows the turbulent kinetic energy distribution of the volute center section and the inner and outer walls of the bridge. The change of turbulent kinetic energy gradient of double volute decreases gradually along the radial direction, and the turbulent kinetic energy gradient near the septum is larger, which indicates that the energy loss is larger. The high-speed fluid in the impeller is mixed with the low-speed fluid in the volute to form a turbulent flow, which causes additional hydraulic loss, resulting in a situation where the turbulent kinetic energy gradient at the inlet of the double volute is locally large. From the whole point of view, the turbulent kinetic energy of the inner and outer wall of the bridge is relatively uniform, and only a relatively large fluctuation occurs at the inlet of the inner wall of the bridge. The main causes of the turbulent kinetic energy changes are the fluctuations of the liquid thrown out by the impeller and the impact of the liquid flow caused by the acceleration turning at the inlet of the diffusion tube.

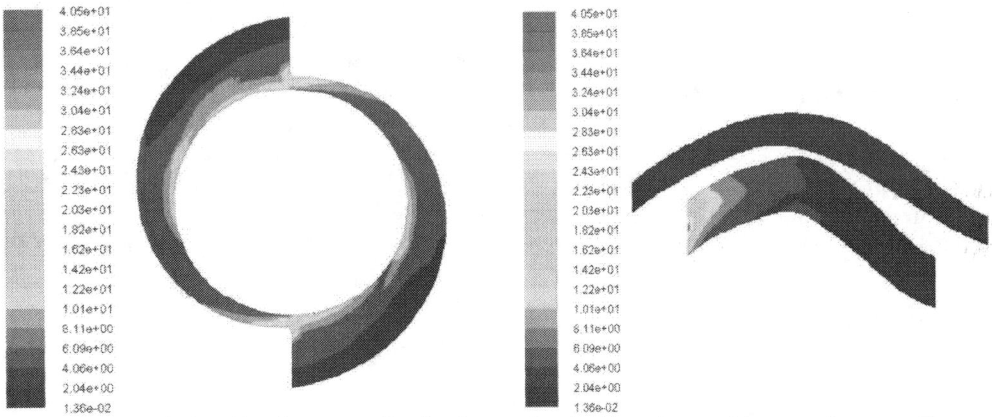

Fig.8. Turbulent kinetic energy distribution on the double volute and inner and outer wall.

Figure 9 shows the velocity vector diagram of the rear section of the bridge, where the speed trajectory is basically consistent with the contour, and the radial trajectory changes along with the speed trajectory. The outer velocity of the guide vane is larger than the inner one, that is, the kinetic energy on the outer side is larger.

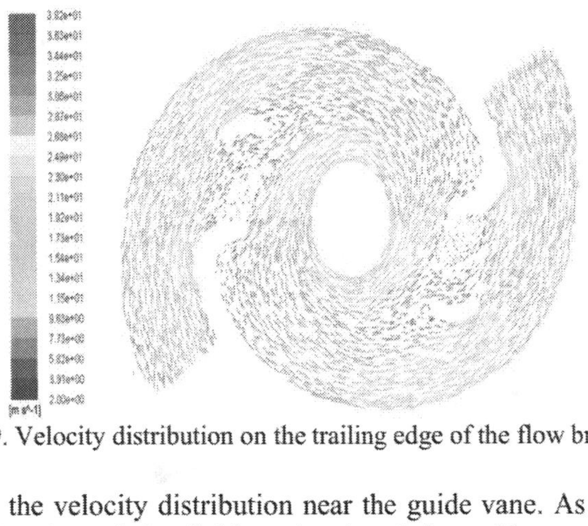
Fig.9. Velocity distribution on the trailing edge of the flow bridge.

Figure 10 shows the velocity distribution near the guide vane. As shown in the figure, the flow velocity direction of the fluid at the tip of the guide vane changes, and flow disorder occurs in a small area, besides, the flow velocity is slightly smaller than the outer side and the velocity vector line is sparse, and the phenomenon of outflow in a small area occurs.

Fig.10. Velocity distribution on the guide vane.

2.2 Structure improvement of multistage middle-open centrifugal pump inter-stage water flow channel

Through the analysis of the inter-stage flow passage, the guide vanes on the bridge are mainly improved to improve the internal flow pattern in the flow passage, and reduce the phenomenon of backflow or vortex in it, reduce energy loss, also improve the performance of the pump. Based on the original design, by changing the angle of the front and rear ends of the guide vane in the horizontal direction [14-15], three different improvements are proposed in table1.

Table 1. The improved schemes of guide vane.

	Front inlet angle	Rear exit angle
Original design	30°	45°
Scheme 1	45°	45°
Scheme 2	60°	30°
Scheme 3	60°	45°

For the above three improved schemes, the numerical simulation was carried out respectively. Through calculating the energy loss of the inter stage flow passage, the one with the least energy loss and the most reasonable design was selected.

2.3 Simulation results analysis of the improved model

In this paper, the multistage middle-open centrifugal pump is used. Its flow channel structure is completely symmetrical from left to right, so the structure of the flow channel between the stages is the same. Therefore, the single-stage channel on both sides is selected for numerical simulation to observe the flow condition near the guide vane, as shown in Figure 11.

Guide vanes are set in the inter-stage flow passage, which are mainly used to introduce liquid into the lower impeller. It can be seen from the flow velocity diagrams of the three improved schemes that the flow velocity in the inner side of the guide vane is less than that in the outer side, the flow velocity direction is almost the same as the shape of the vane, and only a small area of flow disorder appears at the end of the vane. Comparing the three flow velocity diagrams, it can be seen that the disorder areas of the ends of the guide vanes of the improved scheme 2 and the improved scheme 3 are smaller, which is more in line with the flow law.

(a) Improved Scheme 1

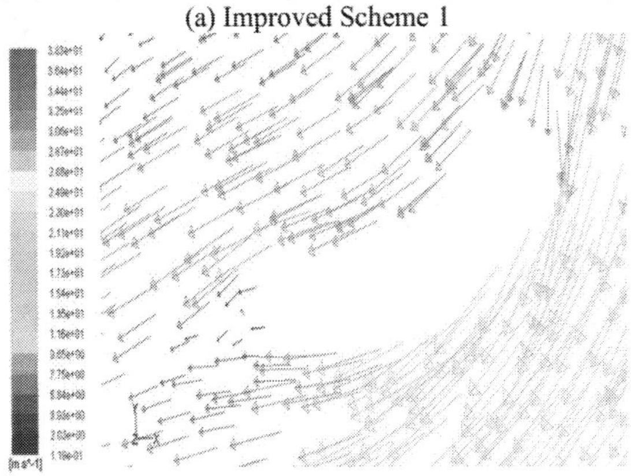

(b) Improved Scheme 2

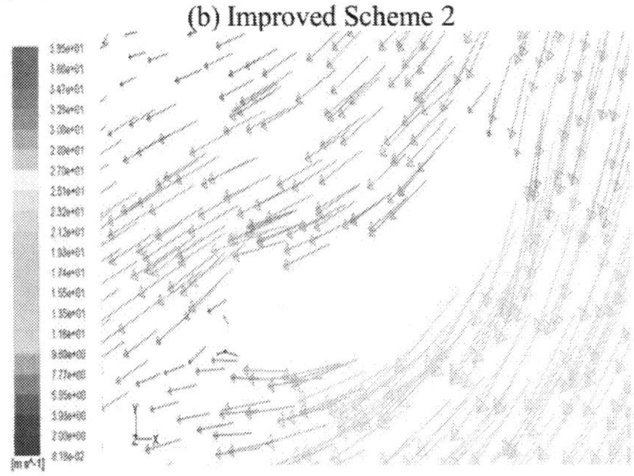

(c) Improved Scheme 3

Fig.11. Velocity distribution near the guide vane.

2.4 Energy loss analysis of four structures

From the perspective of energy loss, the four structures are analyzed quantitatively, and the structure with the smallest loss is finally selected. Since the fluid flows through the inter-stage flow channel inlet and outlet, following the Bernoulli equation, the energy loss at the inlet and outlet can be calculated by Fluent integral. In order to facilitate the analysis of the effect of the change of the guide vane on the performance of the multi-stage pump, the energy loss of the inter-stage water flow channel from the left is h_{right}, and the energy loss of the inter-stage water flow channel from the right is h_{left} and the average of the two is h_{ave}, then the calculation results are shown in Table 2.

Table 2. The energy loss comparison among different designs.

Energy loss	Original design	Scheme 1	Scheme 2	Scheme 3
$h_{left}(m)$	3.276	2.775	2.596	2.879
$h_{right}(m)$	3.273	2.771	2.594	2.873
$H_{ave}(m)$	3.2745	2.773	2.595	2.875

According to table 2, the energy loss of the three improved schemes is less than that of the original design, and the energy loss of scheme 2 is the smallest. Therefore, the structure of scheme 2 is adopted as the final structure for further experimental and comparative research.

2.5 Comparison of calculation and test performance curves

The flow field of multistage pump is simulated under different conditions, and the performance parameters are predicted. The external characteristic test of water pump is carried out on the open test-bed. The main external characteristic curves of multi-stage pumps are: flow-lift characteristic curve, flow-efficiency characteristic curve and flow-axis power characteristic curve. The comparison between the test data and the calculated data is shown in Figures 12, 13, and 14.

From Figure 12, it can be seen that there is a certain error between the test value and the calculated value. The reason may be: under the design condition, there is no obvious backflow and vortex, and the flow field is relatively stable, so the experimental values and the calculated values agree well; while in the non-design conditions, the boundary layer is separated, and there are more local backflow and vortex, so the stability of the flow field is poor, resulting in large errors.

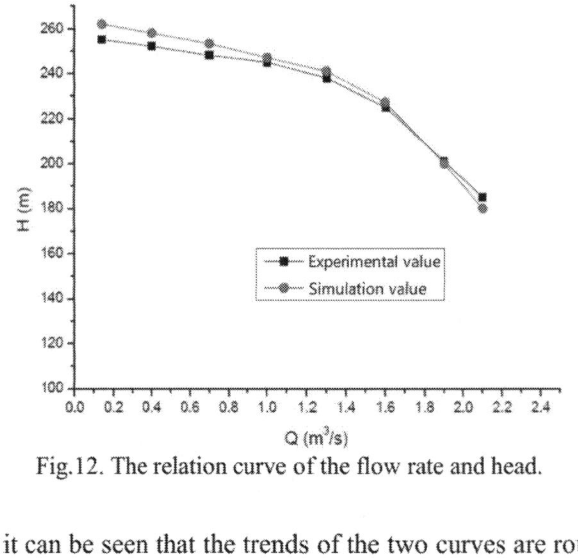

Fig.12. The relation curve of the flow rate and head.

From Figure 13, it can be seen that the trends of the two curves are roughly the same, but the simulation calculation value is on the high side, which may be caused by the simplification of the physical model, and factors such as volume loss and clearance loss are not considered in the simulation calculation.

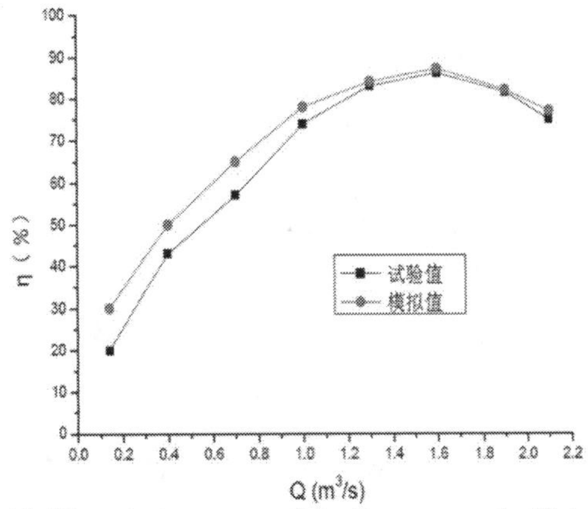

Fig.13. The relation curve of the flow rate and efficiency.

From Figure 14, it can be seen that under different flow conditions, the measured shaft power is slightly larger than the calculated one. The main reason is that the simplified physical model leads to the calculation results not including the power consumed by the loss of volume, the loss of bearing and sealing packing and the friction loss of some disks. In addition, factors such as the low roughness of the solid wall surface defined in the simulation calculation of the internal flow field of the multi-stage pump also cause the shaft power prediction curve to be lower than the test curve.

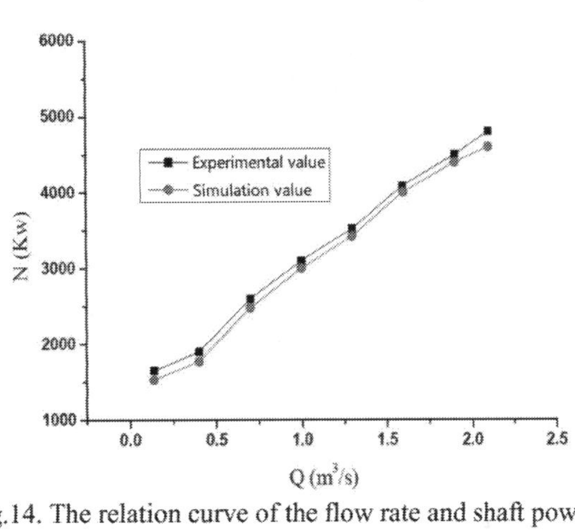

Fig.14. The relation curve of the flow rate and shaft power.

3 conclusions

(1) Based on the performance parameter requirements, this paper designs a hydraulic model of the inter-stage flow channel of the multistage middle-open centrifugal pump, reasonably matches the size of the channel between the stages, and uses Pro/E three-dimensional modeling software for modeling.

(2) The numerical simulation method is used to analyze the flow velocity, pressure and turbulent kinetic energy distribution of the inter-stage flow channel. Simulation results show that the design of the double volute and the tongue is reasonable, and there are backflow and vortices at the ends of the guide vanes passing through the bridge.

(3) Aiming at the flow conditions around the guide vanes of the inter-stage flow passage, based on the original design, three different improvement schemes were proposed and

numerical simulations were performed by changing the horizontal angle of the front and rear ends of the guide vanes, and the numerical simulation was performed to calculate the magnitude of energy loss, comprehensive analysis shows that scheme 2 has the smallest loss. Therefore, it is the optimal hydraulic model.

(4) The steady numerical simulation of the flow field of the multistage middle-open centrifugal pump is carried out under multiple working conditions, combined with experimental verification, shows that the test agrees well with the simulation and the hydraulic design meets the requirements.

References

[1] M. Liu, T. Lei, X. Yun. *J. Petrol. Sci. Eng.* **184** (2020)
[2] F. Lai, Y. Wang, S. A. El-Shahat, G. J. Li. *J. Fluid Eng-T ASME.* **141** (2019)
[3] Shah, I. R. Chuahtai. *Int. J. Multiphase Flow.* **37** (2011)
[4] R. Spence, J. Amaral-Teixeira. *Computers & Fluids.* **37** (2008).
[5] K. Sung, K. Yi, J. H. Kim, Y. S. Choi. *Adv. Mech.* **11** (2019)
[6] Guo, M. Gao, J. Y. Wang, Y. T. Shi, S. Y. He. *Appl. Acoust.* **155** (2019)
[7] Q. R. Si, H. Y. Zhang, G. Bois, J. F. Zhang. *Energies.* **12** (2019)
[8] S. Huang, F. X. Yang, J. Guo. *Procedia Engineering.* **61** (2013)
[9] A. Panda, V. Nahornyi. *Int. J. Adv. Manuf. Tech.* **105** (2019)
[10] X. Q. Jia, Y. Sheng, Z. C. Zhu, B. L. Cui. *Eng. Computations.* (2019)
[11] T. Krzysztof, K. R. Kaczorowska-Ditrich. Flow Tureul. Combust. (2019)
[12] W. Eva, K. Mario, M. Ilona. Ann. Biomed. Eng. **48** (2019)
[13] Y. Yuan, S. Q. Yuan. *Adv. Mech.* **9** (2017)
[14] Y. Wang, B. F. Song, W. P. Song. *J. Vibroeng.* **19** (2017)
[15] J. Shendage, S. B. Kedare, S. L. Bapat. *Energy.* **140** (2017)

Reviews of Research on Mechanical Fault Diagnosis in GIS

Yuqi Cai[1,*], *Liang* Jin[2]

[1]College of Information Science and Engineering, Huaqiao University, 361021,Xiamen, China

[2]College of Information Science and Engineering, Huaqiao University, 361021,Xiamen, China

Abstract. Gas Insulated Switchgear (GIS) has been widely used in recent years due to their advantages of small floor area, high reliability and good safety. It has arised a series of problems. At present, researchers at home and abroad have done more in-depth research on insulation fault, but research on mechanical fault is insufficient. Firstly, this paper reviews the research status of mechanical fault detection technology in GIS at home and abroad. Secondly, the aspect of the principles of mechanical fault in GIS, fault judgment and location technology, and early fault prediction and status evaluation in GIS were summarized. Finally, the paper prospected the future research direction on this basis.

1 Introduction

Gas Insulated Switchgear (GIS), also known as the SF6 fully enclosed combined apparatus, in addition to the transformer, it is all enclosed within a grounded metal shell which is filled with 0.34 MPa to 0.4 MPa SF6 gas as insulation and arcing medium, including circuit breaker, disconnecting switch, grounding switch, current transformer, voltage transformer, lightning arrester, busbar, connecting pipe and transition element[1]. Compared with conventional open type substation, it has many advantages such as small occupied area and space, high operation reliability, good safety, long maintenance period and little influence from external environment, so it is widely used in urban substation [2].

In recent years, the load of GIS equipment of various voltage levels in the power grid has been increasing rapidly, but the resulting problems can not be ignored. According to the statistics of national power reliability, with the continuous improvement of GIS availability coefficient, the forced outage rate has also increased significantly, indicating that the fault problem of GIS has increased to a certain extent [3]. Therefore, compared with conventional electrical equipment, GIS has some disadvantages in operation and maintenance. The main points are as follows: (1) Key components of the equipment are enclosed in the metal shell, and it is very difficult for maintenance personnel to find fault signs. (2) The internal space of GIS is small, but the structure is complex. So the connection between devices is close and the mutual influence is big. It is easy to produce chain reaction by this, increasing the difficulty of fault judgment and positioning. (3) Faults of GIS affects a large range, and it

* Corresponding author: cyq20200213@163.com

takes a long time for check and maintenance, generally more than two weeks [4]. Therefore, more effective on-line monitoring and fault diagnosis techniques are adopted Points is necessary.

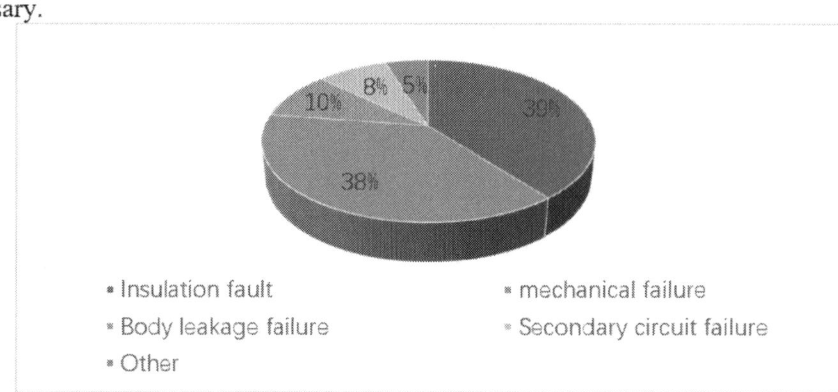

Fig. 1. Fault statistics of GIS

According to various statistical data, GIS faults can be roughly divided into mechanical faults, insulation faults, secondary circuit faults, body leakage faults, other faults, etc[3]. The proportion of each fault is shown in figure 1. As can be seen from figure 1, the main faults of GIS are insulation faults and mechanical faults. Since the birth of GIS in the 1960s, there have been many systematic and in-depth studies on its discharge fault, but the research on mechanical fault is insufficient [4]. Therefore, we need to further study its mechanism and methods of detection and diagnosis.

This paper systematically sorted out the research status of GIS mechanical fault diagnosis at home and abroad, and summarized the characteristics of GIS mechanical fault, fault judgment and positioning technology, GIS early fault prediction and reliability evaluation, and clarified the existing problems in fault detection and future development direction. The overview framework is shown in figure 2.

- *Review framework*
  - *Mechanical fault characteristics of GIS*
  - *Mechanical fault detection of GIS*
    - *Traditional diagnostic methods*
      - *Signal feature extraction*
      - *Fault identification method*
    - *Other diagnostic methods*
  - *Early fault prediction and reliability evaluation of GIS*
  - *Summary and Prospect*

Fig.2. Review structure of Research on Mechanical Fault Diagnosis in GIS

2 Mechanical failure characteristics of GIS

To diagnose GIS mechanical fault, it is necessary to clarify GIS fault mechanism and its typical characteristics. There are three kinds of vibration caused by GIS mechanical failure, including electromagnetic force, electric force and ferromagnetic resonance, Vibration caused by switch operation, and vibration caused by mechanical failure of circuit breaker [6]. Circuit breaker, as an important GIS device, is most prone to mechanical failure. Secondly, dynamic and static contact system and the isolation switch are also common mechanical failure devices. Voltage transformers, electromagnetic contactors and current transformers all work by magnetic flux characteristics, and mechanical and electrical faults often interweave on them.

According to the different vibration frequencies, the GIS state can be distinguished. At present, most researches on GIS vibration frequency are conducted through finite element analysis software simulation or modal test [7,8,9], and some are also conducted through experimental platform for simulation test analysis. The vibration frequencies of different causes obtained from the study are summarized in table 1.

Table 1. Causes and Frequencies of Vibration in GIS.

GIS vibration principle	Frequency characteristics of vibration signals
Partial discharge	The main frequency of vibration is 5-30kHz.
Conductive particles	The vibration frequency is between 1kHz and 10kHz.
Electromagnetic force, electromotive force, electrostatic force	It is mainly 100Hz, and the vibration caused by electric power is very weak when the high-frequency signal is above 300Hz.
Contact surface current shrinkage	The main frequency is 100Hz, and the GIS high voltage conductor has some defects such as poor mechanical contact, which will lead to the increase of vibration signal amplitude, and the increase of signal component amplitude of 200Hz and 300Hz.
Magnetostriction	The fundamental frequency is 100Hz, and the vibration of the core caused by magnetostriction is still very strong at the frequency of 600Hz.
Circuit breaker operation	The vibration frequency range is 100Hz-1kHz.
To ground short circuit	The vibration frequency ranges from 100Hz to 500Hz.

It can be seen from this that the vibration signal can effectively distinguish the mechanical fault of low frequency vibration from the insulation fault of high frequency vibration and the various types of mechanical fault according to its frequency difference.

3 Mechanical fault detection of GIS

The research on fault diagnosis can usually be divided into four levels: first, the existence of fault can be determined; second, the location of fault can be determined; third, the damage degree of fault can be determined; and fourth, residual life prediction and reliability assessment [12]. At present, GIS mechanical fault diagnosis is mainly based on the first three levels. The vibration signal extracted from the operating equipment is used to analyse and identify the fault. The improvement of methods mainly lies in the selection of signal feature extraction and analysis methods.

3.1 Traditional diagnostic methods

Because the vibration signals generated by normal operation or fault of GIS are obvious different from other signals, and the frequency range of various fault types is quite different, most of the mechanical fault detection is based on the extraction and analysis of vibration signals since the birth of GIS fault detection technology.

3.1.1 Signal feature extraction

It is very important for fault diagnosis to extract the effective information from the acquisition signal. The optimization of feature extraction can further improve the accuracy and efficiency of fault diagnosis. At present, the commonly used methods are wavelet transform, fast Fourier transform (FFT), independent component analysis. In [13], a mechanical defect diagnosis method was proposed to analyse abnormal vibration of GIS, and a wavelet noise reduction approach was utilized to process the vibration signal to obtain the corresponding spectrum diagram. By analysing the signal components of each frequency segment in the spectrum, the mechanical fault type of GIS is determined. Similarly, in [14], a GIS mechanical defect diagnosis system and method is proposed, based on independent component analysis and denoising. The independent component method is used to achieve ideal signal-to-noise separation for the collected vibration signals, and the corresponding spectrum diagram is obtained. Then it was matched with the GIS mechanical fault frequency characteristics to determine the fault type. In [15], a GIS mechanical condition detection system based on vibration signals is developed, focusing on the mechanical vibration of TV and busbar faults, which uses traditional wavelets analysis and FFT analysis methods to feature extraction. These are all valid experiments with traditional methods.

Empirical Mode Decomposition (EMD) is an novel signal decomposition algorithm, which is adapted to non-linear and non-stationary signals. Once it was proposed, it attracted widespread attention. Compared with traditional analysis methods, EMD overcomes the limitations of the FFT transform, and unlike the wavelet transform, the wavelet basis function must be selected. It has great time-frequency resolution and adaptability, and can highlight small details in the signals. The main processing steps of EMD are shown in figure 3.

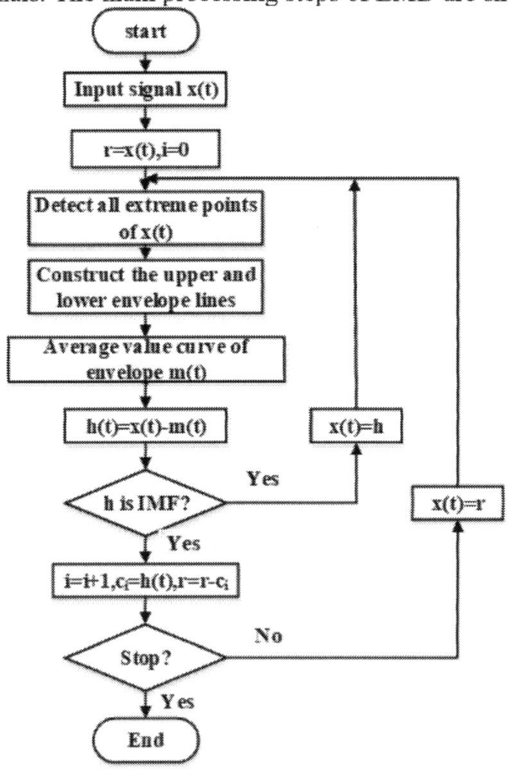

Figure.3. EMD algorithm flow

In [16], empirical mode decomposition is used to extract vibration signals and performed time-frequency analysis; In [17], the researchers apply EMD algorithm to the processing of vibration signals, and the power spectral density function and the vector space model are used to obtain the matrix. The mechanical state of the GIS is monitored by contrasting the power matrix similarity of the vibration signals.

Although empirical mode decomposition is a more mature method of time-frequency signal analysis, there are still modal aliasing phenomena and phenomena that are prone to produce false components. To this problem, in [18], the Ensemble Empirical Mode

Decomposition(EEMD) is used to extract the characteristic amount of the vibration signal, and then the Hilbert transform of the signal obtained through the EEMD is analysed, verified by the GIS vibration signal detection test platform to obtain a more accurate characteristic criterion.

Variational Modal Decomposition (VMD) is an adaptive, non-recursive signal decomposition algorithm based on EMD. It has obvious advantages over EMD and EEMD in terms of signal decomposition accuracy and noise robustness. In [19], a mechanical fault diagnosis method of GIS is published, depended on chaotic valley-buffering algorithm to optimize the variational mode decomposition (VMD) parameters. Finally, VMD entropy, root mean square and the principles of minimum Euclidean distance were utilized to identify the feature vector of samples, and the diagnosis of mechanical fault of GIS is realized.

The above is a more common application of feature extraction technology in GIS machinery fault diagnosis. With its continuous improvement, GIS machinery fault diagnosis technology based on vibration signals will have greater development.

3.1.2 Fault identification method

1) Traditional analysis method

After the fault features are extracted, it is crucial to effectively analyze and match the known fault features. Because the vibration frequency generated by GIS mechanical faults has a large difference and the information displayed in the time-domain waveform of the extracted signal shows less information, the most common method currently is to determine the type of fault through spectrum analysis. In [20], a single-fault experimental study on the looseness of GIS bus bars is carried out. After an in-depth understanding of the fault mechanism, a simulation experiment was performed through the GIS mechanical vibration detection test platform. The fault characteristics were analysed by spectrum. The results show that in addition to the obvious characteristics of the vibration signal at the frequency of 100 Hz, the vibration signal also changes in the frequency range of 1000 to 2000 Hz, especially at the frequency of 1400 Hz. In [13] and [14], after processing the signals, the types of GIS mechanical faults were also determined by analysing the signal components in each frequency band of the spectrogram. In [21], dynamic time warping algorithm (DTW) to perform time-frequency signals is used. Other researchers judged the ratio of vibration amplitudes at different frequencies as characteristic values. After the frequency analysis of the vibration signal in [10], the amplitude ratio between different frequencies is used as the criterion for judging the contact state of the GIS isolating switch. In the same method, the problem of bus looseness is explored in [22].

Compared with spectrum analysis, power spectrum can represent the relationship between signal power and frequency, and its analysis is more effective. In [23], spectrum analysis and power spectrum analysis and diagnosis of GIS vibration signals is carried out at the same time, and then obtained fingerprints through census for comparison to effectively diagnose the fault location. It is confirmed that power spectrum analysis can reflect abnormal states better than spectrum.

Analysing the type of fault by just the frequency spectrum may miss information hard to detect. In [5], the Hilbert-Huang transform is applied to the processing of GIS vibration signals. By comparing the Hilbert spectra of vibration signals with normal operation of GIS and suspected faults, the changes of the amplitude of each frequency component with time were obtained. Then the marginal spectrum of the normal signal and the abnormal signal is compared to get the fault feature quantity of the GIS equipment, so as to determine whether it is faulty. This method has some effects on fault identification, but the fault characteristics displayed by it are not accurate and require a large amount of measured data. So there is space for improvement of this method. Based on this, in [18], EEMD is combined with Hilbert transform to obtain more accurate fault characteristics and improve fault diagnosis efficiency.

2) Fault diagnosis method based on feature classification

As a kind of complex structure equipment, GIS not only contains a lot of electrical devices, but also a large number of mechanical devices. GIS has high reliability, so it is difficult to obtain its fault data, and it often has fewer measured samples. As mentioned above, most methods of GIS mechanical fault diagnosis are verified on experimental test platforms. Because GIS fault detection lacks experience accumulation, algorithms that rely on subjective experience cannot be applied, such as fuzzy theory. The stability of genetic algorithms is poor. The artificial neural network relies heavily on data and is not applicable to small sample problems. However, as a classifier, support vector machine (SVM) has a relatively small empirical risk and confidence range, and can achieve such small sample classification. In the past two years, this method has begun to be applied to GIS mechanical failure. In [24], the combination of wavelet packet decomposition theory and vector machine theory was used to extract fault features and identify the state of high-voltage circuit breakers. Aiming at the problem of high-voltage circuit breaker fault vibration signals caused by insufficient mechanical lubrication, it is difficult to determine the fault if the energy entropy is directly used to extract energy entropy for feature extraction. An energy segmentation method is proposed in [25], such as EMD, to extract feature quantities and combined with support vector machines. Then a method is proposed for diagnosing mechanical faults of high-voltage circuit breakers, verified with common faults. In [26], the amplitude ratios of vibrations at different frequencies are also used as eigenvalues, and then SVM is used for classification and recognition. In [11], VMD energy entropy and root mean square value as the feature values is used, and support vector machine is used ,which is combined with kernel principal component analysis to diagnose. The diagnosis success rate was high.

Aiming at the limitation of SVM in the classification of complex faults, the researchers put forward a failure diagnosis method based on the combination of frequency feature extraction of coherent functions and multi-layer classifiers, which combined two one-class support vector machines (OCSVMS) and one support vector machine (SVM) into a multi-layer classifier. The first OCSVM classifier was utilized to differentiate between normal and abnormal conditions in actual detection. The second OCSVM classifier was utilized to distinguish between known and unknown faults, and solve the problem of error diagnosis of unknown faults. The fault type is determined by support vector machines in failure recognition [27]. This method makes the organic combination of failure judgment, type identification, and condition monitoring to maximize the advantages of SVM.

In addition, there are some classification algorithms that have been used more, such as the principle of cluster analysis. In [28], the neighbourhood algorithm is combined with the fuzzy C-mean algorithm (FCM) to obtain the final classification result to calculate the initial cluster center. Not only can it solve the problem of storage requirements in the neighbourhood algorithm, it has a large amount of calculation and a long calculation time, and it also overcomes the inaccuracy of the cluster center obtained from FCM, which causes the low accuracy of GIS mechanical fault diagnosis.

3.2 Other diagnostic methods

In recent years, with the researchers' emphasis on GIS mechanical failure and the rapid development of GIS online monitoring technology, the judgment of GIS mechanical failure depends not only on vibration signals. Some researchers have used noise signals to judge and locate GIS faults, or comprehensively analyse sound and vibration signals to determine the type of mechanical faults. A self-developed set of front-end abnormal sound online acquisition equipment is introduced in [29], and a signal fault diagnosis algorithm based on an online sequence over-limit learning machine is proposed. The accuracy of the algorithm increases with the increase of the number of GIS sound waves. The device can make up for the blank problems in the field of signal acquisition and online monitoring of early failures of GIS based on acoustic signal analysis. The acoustic characteristics of GIS mechanical faults is summarized in [30], where the time domain, frequency domain and inverse frequency domain laws of different faults are analysed, and a software is designed.

In the future, the means and methods for GIS fault signal collection will become more and more abundant, and the problem of difficult GIS fault signal collection will also be solved.

4 EARLY FAULT PREDICTION AND RELIABILITY EVALUATION OF GIS

The above three are the first three levels of fault diagnosis. The first three levels of fault diagnosis are based on the failure of the equipment, and the fault monitoring and maintenance are performed in time. The purpose of fault prediction is to predict the future health status of the device and the time interval from the current status to the fault status by the prediction method, and to extend the life cycle of the device [31]. Especially for GIS, a highly integrated device with complex internal structure, a series of chain reactions may occur. Once GIS is damaged, it will cause a large range of impacts, make maintenance difficult and take a long time. Faults are divided into complete faults and faint faults according to their development degree. Complete faults refer to equipment failures that have developed to the middle and late stages, and the fault characteristics are obvious. Failure to take measures will cause major accidents; Faint faults usually refer to faint faults or potential faults at early stages or the characteristics of the faults are masked and not obvious. Early prevention of weak faults can avoid the occurrence of complete faults, and mechanical fault diagnosis has gradually shifted from complete fault maintenance to weak fault state evaluation and prediction [12]. A GIS mechanical failure approach method based on KL divergence is put forward in [32]. The kernel density estimation method was used to estimate the probability density of the original measurement signal. The degree of fault occurrence was judged by calculating the divergence between the normal signal and the fault signal. This method can discover potential faults in GIS and locate faults at an early stage.

As a kind of complicated equipment, GIS may cause other faults if one kind of fault occurs. The symptoms it shows are ambiguous and the system needs to sort out the relationship between their failures. The GIS fault tree is established in [3], and the logical relationship between GIS faults is sorted out. Then the reliability of GIS is evaluated. A more in-depth analysis and research on the causes of GIS failures has shown that the failure of manufacturing quality and the lack of strict control over the installation are the main reasons for GIS failures. In [33], Petri net theory is organically combined with fuzzy inference rules, and fuzzy Petri net (FPN) is proposed, which has fast and accurate parallel reasoning capabilities. It uses historical fault data and FPN forward matrix reasoning to perform reliability analysis on GIS. In the case of a known fault phenomenon, the cause of the fault is found out by the reverse fuzzy reasoning rule. Compared with the fault tree, this method has faster and more accurate reliability analysis and fault diagnosis, but it requires a large number of statistical fault cases in order to establish a more comprehensive fault diagnosis model. In [34], according to the actual functional requirements of GIS condition monitoring, a GIS intelligent monitoring status information flow modeling method in accordance with the IEC 61850 communication standard was proposed , and a GIS condition monitoring IED model was established. And the architecture and communication system of GIS intelligent monitoring system are given. In the future, substations will become more intelligent, and data sharing will become more and more convenient. Intelligent diagnosis systems that integrate monitoring and early warning, intelligent diagnosis, and condition assessment will become an inevitable trend.

At present, it is still difficult to obtain the entire process data of the state change of GIS mechanical failure, which requires us to make a prejudgment of the monitoring status based on the effective learning of normal operation data.

5 SUMMARY AND PROSPECT

In recent years, as one of the main failures of GIS, mechanical failure has attracted more and more attention from researchers. The principle of GIS mechanical failure and its detection technology have been further developed. This paper systematically summarizes the research status of GIS mechanical fault diagnosis technology at home and abroad, and summarizes the characteristics of GIS mechanical faults, fault judgment and location technology, early fault prediction and reliability evaluation of GIS.

1) The research of GIS mechanical faults mostly focuses on single faults, and there are few researches on the types of faults that coexist with multiple mechanical faults. This requires in-depth research on multi-feature signal extraction and processing techniques. Moreover, the verification of diagnostic methods is mostly performed on a simulation test platform, and the effect of actual situation verification is difficult to guarantee.

2) Due to the difficulty of GIS maintenance and the large impact of faults, the research on GIS condition assessment is very necessary, but there are few studies in this area.

3) There is no systematic establishment of a full database of time-frequency maps and acoustic characteristics of various types of GIS faults, which is not conducive to ensuring the accuracy and efficiency of GIS mechanical fault diagnosis.

The above is a summary of the problems at this stage. I hope this article can provide directions for future research on GIS mechanical fault diagnosis.

ACKNOWLEDGEMENT

This work was supported by the Subsidized Project for Postgraduates' Innovative Fund in Scientific Research of Huaqiao of University.

REFERENCES

1. Wang Changchang, Li Fuqi, Gao Sheng you. On-line Monitoring and Diagnosis for Power Equipment[M]. Beijing: Tsinghua University Press, 2006:252-254.
2. Zuo Yafang.Operation maintenance and troubleshooting of GIS. Beijing: China Electric Power Press, 2013: 12-15.
3. Cui Yangliu, Ma Hongzhong, Wang Taoyun,et al. Fault Analysis of GIS Based on Fault Tree Theory[J]. High Voltage Apparatus,2015(7):125-129.
4. Sun Q, Cao T, Hou Y, et al. Detection and Analysis Based on the Abnormal Mechanical Vibration Signal of GIS[C]. FifthInternational Conference on Instrumentation and Measurement Computer, Communication and Control (IMCCC). Qinhuangdao, China: IEEE, 2016.
5. Xu Tianle, Ma Hongzhong, Chen Kai, et al. Diagnosis of GIS Failure Based on HHT Method for Vibration Signals[J]. Electric Power,2013, 46(3):39-42.
6. Chen Xuanhong, Gaoyang, Xu Aolan. Mechanical failure principle of GIS equipment [J]. Shandong Industrial Technology, 2017 (19): 162-162.

7. Li Kai, Xu Honghua, Chen Bingbing,et al.Research on vibration mechanism and natural frequency in GIS[J]. Electrical Measurement and Instrumentation, 2017, 54(3),14-18.
8. Yang Jinggang, Liu Yuan, Song Siqi, et al.Research on Vibration Detection Technology of GIS Mechanical Defects[J].High Voltage Apparatus,2018, 54(11):94-98.
9. Qi Weidong, Niubo, Hu Degui, et al. Vibration Simulation of GIS Horizontal Bus Shell Based on FEM[J]. High Voltage Apparatus,2018,54(6):46-59.
10. Jia Yongyong, Wu Bang, Shan'yuan Sun, et al. Research of the vibration characteristics in GIS Disconnector under different Contact State[C]. 2018 IEEE International Conference on High Voltage Engineering and Application (ICHVE), ATHENS, Greece,2018.
11. He Siyang, Zhang Tangsong, Zhou Lan,et al.GIS fault diagnosis strategy based on feature classification algorithm[J].Automation and Instruments,2019, 232(02):203-206.
12. Wang Guobiao, He Zhengjia, Chen Xuefeng, et al.Basic Research on Machinery Fault Diagnosis—What is the Prescription[J].Journal of Mechanical Engineering,2013, 49(1):63-72.
13. Sun Qingsheng ; Cao Tao,et al.Detection and Analysis Based on the Abnormal Mechanical Vibration Signal of GIS[C].Proceedings of 5th International Conference on Instrumentation and Measurement, Computer, Communication, and Control, Qinhuangdao, China, 2015.
14. State Grid Corporation. GIS Machinery Defect Diagnosis System and Method Based on Independent Component Analysis Denoising: China, 201510971975.3 [P]. 2016-06-01.
15. Huang Jinjian,Yang Baokun.GIS Mechanical Condition Detection System Based on Vibration Signaland Its Application[J].Guangxi Electric Power, 2015, 38(5):55-59.
16. Sun Qingsheng, Cao Tao, Zhou Zhangbin, et al. Mechanical Vibration Signal Analysis of GIS Based on EMD Algorithms [J]. Electrical Applications, 2016 (20): 79-84.
17. Zhigang Wu, Bing Liu, Lijuan Wu , et al. Analysis of GIS Mechanical Vibration Signal Based on EMD Algorithm[C]. 3rd International Conference on Information Science and Control Engineering (ICISCE),Beijing, China: IEEE , 2016.
18. Zheng Hao, Zhu Shenglong, Ouyang Yu, et al. Time – frequency Analysis of GIS Mechanical Vibration Signals Based on Improved HHT Algorithm[J]. Computer Technology and Applications,2018,40(1):15-18.
19. Gao Xiaodong, Lv Shouguo, Zhuang Yanfei, et al. Method for Diagnosis of Gas Insulated Switchgear Based on VMD Parameters Optimized by Chaotic Cuckoo Algorithm[J]. 13th IEEE Conference on Industrial Electronics and Applications (ICIEA), Wuhan, China,2018.
20. Huang Qing, Wei Xu, Xu Jiangang, et al.Research of GIS Bus Loose Contact Fault Diagnosis Techniques Based on Mechanical Vibration[J].High Voltage Apparatus,2017,53(11):99-108.
21. Michel Landry, FranÇois Leonard , Champlain Landry, et al. An Improved Vibration Analysis Algorithm as a Diagnostic Tool for Detecting Mechanical Anomalies on Power Circuit Breakers[J]. IEEE Transactions on Power Delivery, 2008, 23(4):1986 - 1994.
22. Liu Yuan , Jia Yongyong, Yang Jinggang. Research of Mechanical State Diagnosis Techniques in GIS Bus Connector Based on Mechanical Vibration[C]. 12th International Conference on the Properties and Applications of Dielectric Materials (ICPADM), Xi'an, China,2018.

23. Cheng Lin.Research on UHV GIS/HGIS Vibration Detection and Diagnosis Method[J].Electric Power Construction,2009,30 (7):17-19.

24. Sun Laijun.A New Method of Fault Diagnosis for High Voltage Circuit Breakers Based on Vibration Signals[J].Proceedings of the CSEE, 2006, 26(6): 157-161.

25. Huang Jian, Hu Xiaoguang, Gong Yunan.Machinery Fault Diagnosis of High Voltage Circuit Breaker Based on Empirical Mode Decomposition[J].Proceedings of the CSEE,2011, 31(12): 108-113.

26. Feng Ying, Lan Jian, et al. Bolt looseness fault diagnosis of GIS based on vibration signal amplitude[C]. International Conference on Mechanical, Electric and Industrial Engineering, Hangzhou, China,2018.

27. Yuan Yang, Ma Suliang, Wu Jianwen, et al. Frequency Feature Learning from Vibration Information of GIS for Mechanical Fault Detection[J]. Sensors, 2019, 19(8): 1949.

28. Shen X, Lin Z, Peng G, et al. Research on mechanical fault diagnosis of Ultra high voltage GIS based on the combination of neighbour algorithm and FCM[C]. IEEE PES Asia-Pacific Power and Energy Engineering Conference (APPEEC), Xi'an, China, 2016.

29. Wang Zhe, Gao Wei, Yang Fan, Leiyuan. An On-line abnormal sound monitoring device for the diagnosis and location of mechanical defects in GIS ontology[J]. Electronic measurement technology, 41 (21): 103-108.

30. Shang Yu,Liu Qiang ,Niu Bo ,et al. Mechanical Fault Diagnosis System Based on Acoustic Feature Analysis in Gas Insulated Switchgear[C]. 1st International Conference on Electrical Materials and Power Equipment (ICEMPE), Xi'an, China,

31. Xu Yuanjun, Pan Hongxia.Study on Tendency Prediction in Equipment Failure[J]. Journal of Vibration,Measurement and Diagnosis, 2006, 26(4): 305-308.

32. Hou Yan,Zhao Tong,Hao Zheng,et al.Discriminant Approach to the GIS Mechanical Fault Diagnosis Based on the K-L Divergence of Vibration Signals[C].IEEE International Conference on High Voltage Engineering and Application (ICHVE),Chengdu, China,2016.

33. Wang Taoyun, Ma Hongzhong, Cui Yangliu, et al. Fault diagnosis and reliability analysis for GIS based on fuzzy Petri net[J].Advanced Technology of Electrical Engineering and Energy, 2016, 35(5):67-73.

34. Du Fengqing, Sheng Gehao, Xu Jian, et al. Information modeling for GIS smart monitoring based on IEC61850 and development of information interaction system[J]. Electric Power Automation Equipment, 2013, 33(6): 163-167.

A Review of Fault Diagnosis Research on On-Load Tap-Changers

Liang Jin[1,*], *Yuqi* Cai[2]

[1]College of Information Science and Engineering, Huaqiao University, 361021 Xiamen, China

[2]College of Information Science and Engineering, Huaqiao University, 361021 Xiamen, China

Abstract. As the core component of transformer voltage regulation, on-load tap changer plays an important role in ensuring voltage stability of power system. With the increase of times of voltage regulation, the failure rate also increases correspondingly. In view of the common faults of on-load tap changer, this paper introduces several existing fault diagnosis methods and summarizes their advantages and disadvantages. Finally, the problems in fault diagnosis of on-load tap-changer and the future research focus are analysed.

1 Introduction

An on-load tap-changer (OLTC) is an essential part of a power transformer. It adjusts the output voltage of the transformer by changing the tap position to respond to changes in power generation and consumption. As an important voltage regulating equipment in the power system, the on-load tap-changer can automatically change the taps, adjust the transformation ratio, adjust the voltage and reduce the loss during the system operation, so it has been widely used. Through investigation and statistics of OLTC failure modes, it is found that transformer failures caused by OLTC irregular operation account for more than 27% of the total number of transformer accidents, which is the main reason for transformer failure [1-3]. This article will summarize the research methods of OLTC fault diagnosis, and put forward the remaining problems and analysis of the links that can be broken.

2 The extraction method of fault feature

Since mechanical failure is the main fault of OLTC, and vibration signal contains a large amount of equipment status information, the mainstream method is based on the feature extraction of vibration signal. At present, the research focuses on vibration feature extraction and multi-fault diagnosis

2.1 Wavelet decomposition

* Corresponding author: 15556921897@163.com

Signal denoising analysis is a pre-processing method that must be done before fault feature extraction. Here are a few simple steps [4]:

(1) Wavelet decomposition. Calculate the high-frequency coefficients and low-frequency coefficients of each scale according to the determined optimal mother wavelet base and the optimal number of decomposition layers. The calculation formula is as follows:

$$\alpha_{j,k} = < S(i), \varphi_{j,k}(\mathrm{i}) > \tag{1}$$

$$d_{j,k} = < S(i), \psi_{j,k}(\mathrm{i}) > \tag{2}$$

Where j is the Number of layers of wavelet decomposition; $S(i)$ is the noisy partial discharge signal; $\varphi(i)$ is scale function used to calculate high frequency coefficients; $\psi(i)$ is wavelet function used to calculate low frequency coefficients.

(2) Wavelet reconstruction. Reconstructing each layer of high-frequency signals Sdj and the highest-scale low-frequency signals $Saj_{\max}$ using the high-frequency coefficients of each layer and the highest-scale low-frequency coefficients.

(3) Threshold denoising of signals at each scale. Threshold denoising is performed on the obtained high-frequency signals of each layer and the low-frequency signals of the highest scale, and the denoised signals are simply added to obtain the denoised waveform $S'(i)$.

In view of the non-stationary and nonlinear characteristics of vibration signals, wavelet has its own defects in fault signal decomposition, and its decomposition effect depends on the selection of wavelet basis and decomposition scale, so it is not adaptive. In addition, there is energy leakage in wavelet decomposition. Empirical mode decomposition (EMD) has been widely used in the analysis of vibration and other non-stationary signals because of its adaptive time-frequency analysis.

2.2 Empirical mode decomposition

The empirical mode decomposition (EMD) method is an adaptive signal time-frequency processing method proposed by NASA Huang et al [5-6]. It has proven to be very versatile in a wide range of applications for extracting signals from data generated in noisy nonlinear and non-stationary processes. For a given signal $x(\mathrm{t})$, the EMD decomposition processes are as follows [7] :

(1) primary selection: According to the extreme value of the signal waveform, the upper and lower envelopes are obtained by cubic spline interpolation with the maximum and minimum values, and the average m(t) of the two envelopes is obtained. And remove the average in the signal, as the initial value $\mathrm{h}_1(\mathrm{t})$ of the intrinsic mode function (IMF).

$$\mathrm{h}_1(\mathrm{t}) = \mathrm{x}(\mathrm{t}) - \mathrm{m}(\mathrm{t}) \tag{3}$$

(2) check: check if $h_1(\mathrm{t})$ is an IMF. If not, continue to perform the first screening process on $h_1(\mathrm{t})$ until the obtained component can be satisfied, then define it as IMF and record it as F_1.

3) recycle: take as the new signal, repeat the previous two steps until all IMF components are decomposed, and define the res decomposition margin, then the original signal can be re-expressed as:

$$x(t)=\sum_{i=1}^{n}F_i+res \tag{4}$$

Although EMD has good adaptability and completeness, its modal aliasing problem is always unavoidable. Modal aliasing will not only cause distortion of the time-frequency distribution, but also make the physical meaning of the intrinsic mode function ambiguous [8-9].

2.3 Ensemble empirical mode decomposition

Ensemble empirical mode decomposition (EEMD) according to the statistical characteristics of Gaussian white noise, that is, a unified frequency distribution, white noise is added to the original signal, and then it is subjected to multiple EMD decompositions. The average value of the IMF component is finally used as the final result. The general steps are as follows [10]:

(1) Add white noise to the original signal $x(t)$ to get a new signal $x_j(t)$,then

$$x_j(t)=x(t)+\varepsilon s_j(t) \tag{5}$$

Where $s_j(t)$ is the J-th addition of the white noise sequence, ε is amplitude factor of the added noise;

(2) EMD $x_j(t)$ to get a set of i-layer IMF;

(3) Repeat steps 1 and 2 Ne times, and each time the white noise sequence is different;

(4) Average all IMFs by layer, that is:

$$\overline{c_i}(t)=\frac{1}{N_e}\sum_{j=1}^{N_e}c_{i,j}(t) \tag{6}$$

Where $c_{i,j}(t)$ is the i-th layer IMF obtained at the j-th time;

5) Finally, the decomposition result of is:

$$x(t)=\sum_i\overline{c_i}(t)+\overline{r}(t) \tag{7}$$

EEMD can effectively suppress mode aliasing and extract fault characteristics of vibration signals, but at the cost of computational cost. In addition, the added white noise amplitude and iteration times depend on human experience setting, and modal aliasing cannot be overcome when the value setting is not correct[11-12]. In order to find a more effective OLTC online monitoring and state assessment method that can be used in engineering practice, phase space reconstruction theory can be used to research the chaotic characteristics of vibration signals.

2.4 Phase space reconstruction

The theory of phase space reconstruction was first proposed by Packard [13] and Takens [14] to study the chaotic characteristics of nonlinear signals. Through theoretical analysis and numerical calculation, it is proved that the phase space reconstruction can keep the geometric invariance of the original nonlinear dynamic system, which lays the foundation for the study of high-order nonlinear dynamic system by experimental method. As for the application of phase space reconstruction in OLTC fault diagnosis, a method based on multi-dimensional coordinate space is proposed to extract the geometric characteristics of the phase orbit diagram. By analysing the dynamic characteristics of the mechanical vibration signal of the on-load tap-changer, the specific indicators reflecting the mechanical performance are extracted. However, the phase distribution coefficient used ignores the vector characteristics of the reconstructed signal in the high-dimensional space, and can only express the relative density of the reconstructed signal., so the ability of pattern recognition for the vibration signal is limited, and there may be some unrecognized vibration signals and mis-judgments. In order to solve the above problems, some scholars use k-means clustering analysis method to identify the reconstructed vibration signal mode, and then monitor the mechanical performance of on load tap changer. The research shows that the phase trajectory obtained by the method contains abundant vibration mode information, which makes up for the difficulty of modelling the mechanical characteristics of OLTC, and provides a richer information source for the mechanical condition monitoring of OLTC. However, this method needs to establish a complete vibration fingerprint database, which can provide a basis for OLTC state evaluation and state maintenance. In addition, it is also necessary to continuously accumulate field monitoring data, analyse the changes of cluster center vector sum under the abnormal operation and typical faults of OLTC switch. Relatively speaking, the workload in the early stage will be very large. Some researchers are thinking about whether the operation state of OLTC can be judged by acoustic vibration.

2.5 Acoustic vibration

There are few researches on the modelling of OLTC vibration theory, feature extraction and fault diagnosis of mechanical state. Some researchers have put forward an OLTC mechanical state feature diagnosis method based on the acoustic vibration analysis technology. The vibration model of tap changer is established by the finite element analysis method. The vibration signal collected is denoised by the notch method. The vibration waveform of the switching process is intercepted by the short-time energy method. The characteristic value of the waveform is extracted by STFT based on the Blakeman window. The characteristic value data is obtained by the eigenvalue data. The reference threshold of OLTC in normal state is obtained and OLTC fault is identified. The steps of fault diagnosis based on audio are as follows[15]; first, extract useful data, normalize the amplitude of collected signal, and calculate the correlation coefficient of each tap; then resample the data, two different recording systems are used to record audio and vibration fingerprint at the same time; in addition, align the audio signal and vibration signal, and extract its envelope, that is, signal pre-processing Finally, the correlation coefficient is calculated to determine the relationship between the audio signal and the vibration signal. It should be noted that: 1. The original signal should be pre-processed (such as signal envelope extraction); 2. Hilbert transform acts as a low-pass filter to eliminate the fast oscillation. In addition, the Hilbert transformation can be reduced because the high frequency component in the untreated signal will cause the deviation between samples, otherwise the calculated Pearson coefficient value is much lower than the Pearson coefficient value in the signal envelope, which may lead to the false conclusion.The biggest advantage of the fault detection method based on the acoustic vibration signal is that it can monitor on-line without shutting down the transformer, and can carry out the fault diagnosis of the tap changer when the transformer is live. However, the fault type can't be determined, and the acoustic emission signals need to be collected continuously, and additional devices are needed, which not only increases the monitoring cost, but also takes up more space. In addition, the interference of the environment sound and the sound from the nearby machine to the audio signal is also inevitable, which needs further research on the signal acquisition device.

3 OLTC comprehensive fault diagnosis

The fault diagnosis expert system for OLTC mechanical properties is also applied, as shown in Figure 1. Because the diagnosis results of the on-load tap-changer fault diagnosis expert system depend on the fault symptoms, and the premise of accurately extracting the fault symptoms is that the on-line monitoring system captures the multi-channel sensor measurement signals without omission and carries out reasonable signal processing. Therefore, in addition to the artificial intelligence and calculation methods involved in the system development process, the dynamic detection closely related to the expert system. The design of data acquisition, signal analysis and other systems is also the content of fault diagnosis expert system in a broad sense.

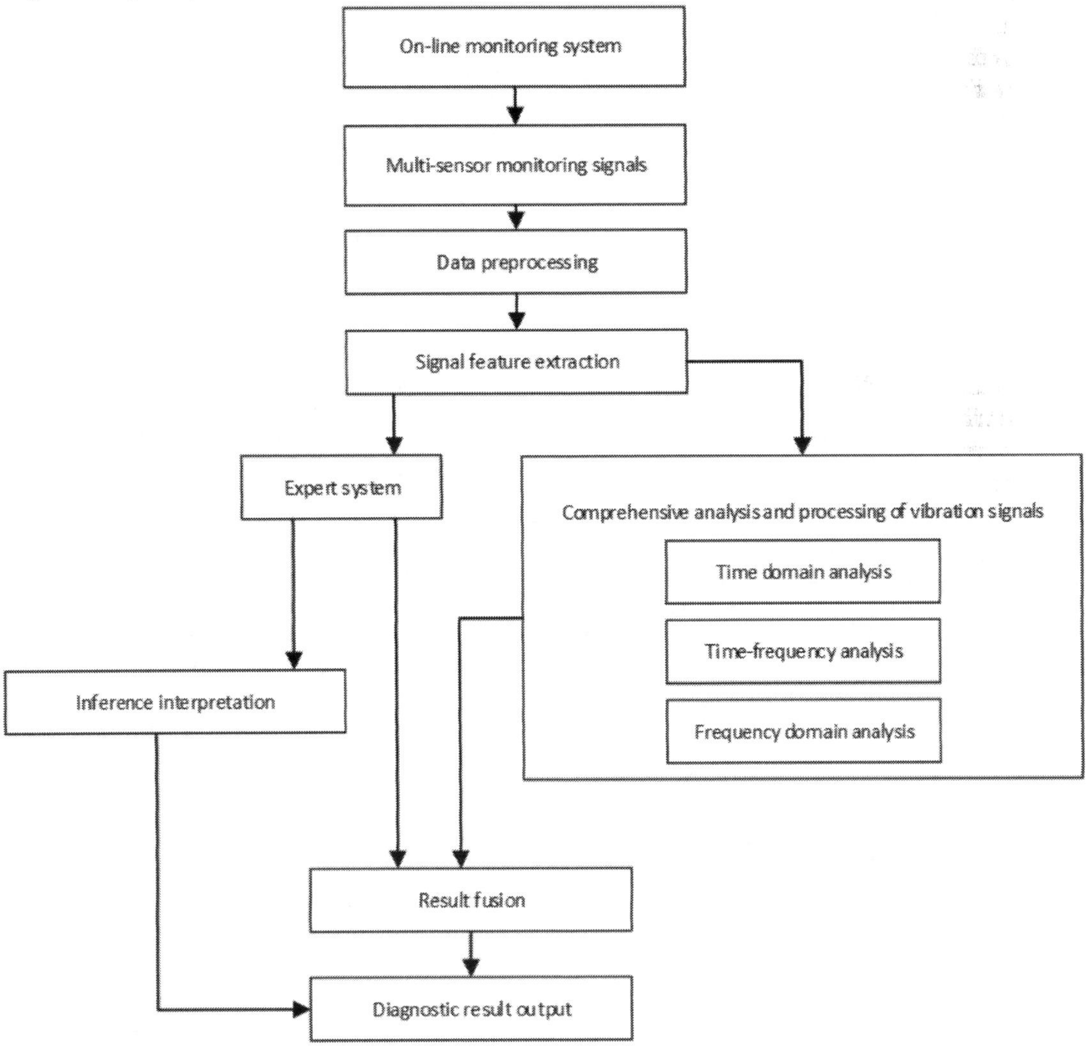

Fig. 1. Overall Structure of the Integrated Fault Diagnosis System.

The above system can make good use of the advantages of decision tree, which is good at dealing with non-numerical data and reasoning quickly, and can also give full play to the multi angle and comprehensive analysis and diagnosis ability of vibration signal comprehensive diagnosis program. But the fault diagnosis object is only for vibration signal. However, although the fault of on load tap changer is mainly mechanical fault, there are still other faults mentioned in the first section, that is, the above system diagnosis object is not comprehensive. Therefore, in addition to the qualitative analysis of oil service test and the fault location analysis of electrical parameters, the comprehensive diagnosis of tap changer fault must also be carried out scientifically and orderly by taking various effective methods and approaches such as targeted gas chromatography analysis and electrical detection means in combination with the operation, historical data and fault characteristics of the tap changer break.

4 Conclusion

In this paper, several effective fault feature extraction methods are summarized, and their advantages and disadvantages are analyzed. In the mechanical fault diagnosis, the general process is shown in Figure 2:

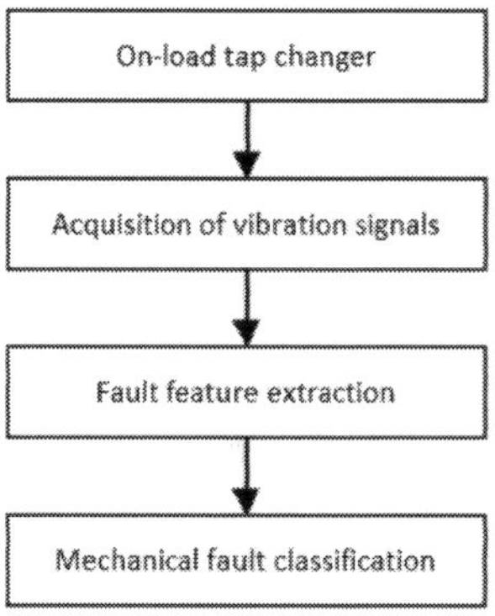

Fig. 2. Mechanical Fault Diagnosis Flow Chart with On-Load Tap-Switch.

So in the diagnosis process, we need to pay attention to the three key steps shown in Figure 2. In the process of vibration signal acquisition, attention should be paid to the position and number of sensors. If necessary, the surrounding noise should be shielded, especially based on the acoustic vibration analysis. Under the influence of the bad working environment on site, considering the anti-interference of the signal, how to design the circuit, match the amplification circuit, filter the clutter through the high-performance digital filter after the signal filtering and analog-to-digital conversion remains to be studied [16]. In the process of fault feature extraction, different methods should be selected for different signals. There are many kinds of faults in on load tap changer, and most of them are mixed faults. Different parts and different types of faults lead to different changes in the functions of the tap changer, resulting in different changes in the operation parameters of the overall and various parts of the tap changer. At present, the diagnosis method is still offline. The problems of online monitoring, the development of signal acquisition system, signal processing technology and the optimal selection of vibration measurement points need further research.

ACKNOWLEDGEMENT

This work is supported by the Subsidized Project for Postgraduates' Innovative Fund in Scientific Research of Huaqiao of University.

References

1. J. Seo, H. Ma and T. K. Saha, IEEE Transactions on Power Delivery, vol. **32**, no. 2, pp. 1031-1038, April 2017.
2. Liu JX , Wang G , Zhao T and Zhang L . ENERGIES, 2017, **10**(7); 946 . 1-14
3. J. Seo, H. Ma and T. K. Saha, IEEE Transactions on Power Delivery, vol. **33**, no. 4, pp. 1689-1698, Aug. 2018.
4. L. Jae-Yoon, L. Dae-Jong, and J. Pyeong-Shik, Trans. Korean Inst. Elect. Eng. P, vol. **65**, no. 3, pp. 188–193,2016
5. J. C. Castro, G. S. Lagos and O. A. Gonzalez, IEEE Latin America Transactions, vol. **15**, no. 10, pp. 1901-1907, Oct. 2017.
6. Q. Li, T. Zhao, L. Zhang and J. Lou, IEEE Transactions on Power Delivery, vol. **27**, no. 2, pp. 596-601, April 2012.
7. Coifman R R , Meyer Y , Wickerhauser V . Wavelets & Their Applications, 1992, **9**(3):765–780.
8. Huang N E. Huang, N.E. et al. Proceedings Mathematical Physical & Engineering Sciences, 1998, **454**(1971):903-995.
9. Yao J , Xiang Y , Qian S , et al. Noise source identification of diesel engine based on variational mode decomposition and robust independent component analysis[J]. Applied Acoustics, 2017, **116**:184-194.
10. Wu Z , Huang N E . Advances in Adaptive Data Analysis, 2009, **01**(01):1-41.
11. Packard N H, Crutchfield J P, Farmer J D, et al. Phys. Rev. Lett., 1980, **45**: 712-715.
12. Takens F. Berlin: Springer, 1981.
13. Majchrzak H Cichon A and Borucki S.ARCHIVES OF ACOUSTICS, 2017, 42(1); Vol. **42**, No. 1, pp. 29–35
14. A. Secic and I. Kuzle, 2018 International Symposium on Industrial Electronics (INDEL), Banja Luka, Bosnia and Herzegovina, 2018, pp. 1-6.
15. Z. Li, Q. Li, Z. Wu, J. Yu and R. Zheng, IEEE Access, vol. **6**, pp. 24148-24156, 2018.
16. Roncero-Clemente C,Roanes-Lozano E.Applied Mathematics and Computation, 2017:S0096300317301285.

AUTHOR INDEX